Quantum Mechanics for Hamiltonians Defined as Quadratic Forms

Quantum Mechanics for Hamiltonians Defined as Quadratic Forms

by

Barry Simon

Princeton Series in Physics

Princeton University Press

Princeton, New Jersey · 1971

L.C. Card: 73-146648
I.S.B.N.: 0-691-08090-9

Printed in the United States of America
by Princeton University Press

To my Parents

INTRODUCTION

It is our purpose in this monograph to present a complete, rigorous mathematical treatment of two body quantum mechanics for a wider class of potentials than is normally treated in the literature. At the same time, we will review the theory of the "usual" Kato classes, although no attempt has been made to make this review exhaustive or complete. The scope of what we present is best delineated by stating the limits of this work: we take for granted the standard Hilbert space formalism, and our main goal is to prove forward dispersion relations from first principles. For example we do not assume the Lippman-Schwinger equation but prove it within the framework of time-dependent scattering theory.

Mathematical physics of the type we discuss here seems to me to be of interest to the wider physics community for a variety of reasons:

(a) It can often clarify concepts considerably. On the pedagogic level, we might mention the relation between boundary conditions and self-adjointness; this connection should help students understand one of those points most often found puzzling to the novice, the seemingly ad hoc introduction of boundary conditions. On the professional level, we might mention the Jauch-Cook work [24, 61] on scattering: the more mathematical approach cleared the cobwebs considerably, eliminating the nonsense of adiabatic switching, i.e., the replacement of V with $Ve^{-\varepsilon|t|}$ and the taking of the $\lim_{\varepsilon \downarrow 0} S(\varepsilon)$. (One can't help picturing a pair of experimenters in the lab, one shouting to the other: "I have the beam ready, turn on the target but do it very slowly!")

(b) It can uncover some interesting subtleties. The standard paradigm which comes to mind is the Wigner-von Neumann example [93] of a potential with a positive energy bound state.

(c) It can often serve as a guide to other fields, mostly notably high energy physics. The classic example is Regge's famous work on analytic continuation in angular momentum. (It must be mentioned that Regge's work came as an outgrowth of his study of the Mandelstam conjecture for Yukawian potentials, a problem suggested by high energy physics.)

(d) Finally, on a certain philosophical level, it is of use to know exactly which "reasonableness" arguments are really true and which require additional conditions.

There is a standard and highly developed lore, most often associated with the work and name of T. Kato, which studies the quantum mechanics of four closely related classes of potentials:

$$L^2 + L^\infty,\ L^2 + (L^\infty)_\varepsilon,\ L^2 \text{ and } L^2 \cap L^1,$$

$$[L^2 + (L^\infty)_\varepsilon = \{V \mid (\forall\varepsilon)\, V = V_{1\varepsilon} + V_2;\ V_{1\varepsilon} \in L^2,\ \|V_{2\varepsilon}\|_\infty < \varepsilon\}].$$

That is, the largest body of results (at least in R^3!) is stated for potentials of these classes (which we call the L^2 classes) and a more or less satisfactory theory exists for each of them, at least in the two body-case. While there are isolated results on more general classes of potentials (e.g., [29, 70, 78]), there does not appear to be any systematic treatment of a larger class.

Before discussing in detail the exact classes we will consider, let us discuss the physical reason that suggests such a wider theory exists. Consider a potential $V(r) = -r^{-\alpha}$. This stops being in any of the L^2 classes when $\alpha = 3/2$ (i.e. $V \in L^2 + (L^\infty)_\varepsilon$ if and only if $0 < \alpha < 3/2$). But physically, it is only when $\alpha = 2$ that one expects bad things to happen — at this point, for example, the centrifugal barrier stops dominating the small r behavior and it is at this point that the freshman physics proof of the lower boundedness of the energy breaks down. (This is the argument that says $V \sim -(\Delta r)^{-\alpha}$ and $T \sim \hbar^2/2m(\Delta r)^2$ by the uncertainty principle.) We thus demand that our larger class include the $r^{-\alpha}$ $(\frac{3}{2} < \alpha < 2)$ potentials.

We will deal specifically with four classes:

$R + L^\infty$, $R + (L^\infty)_\varepsilon$, R, $L^1 \cap R$, where R is the family of potentials obeying:

$$\int \frac{|V(x)|\ |V(y)|}{|x-y|^2}\, d^3x\, d^3y < \infty .$$

The naturalness of this condition is perhaps best demonstrated by mentioning that it appears to have been independently isolated by Rollnik [98], Schwinger [101], Grossman-Wu [44, 45] and Scadron-Weinberg-Wright [100]. There has been some study of this condition by Coester [21], Kato [70], Kuroda [80, 82] and Ghirardi-Rimini [42]. What none of these authors have attempted is a study of potentials of these classes from first principles; in fact, when Grossman-Wu worried about first principles, they added the condition $V \in L^2$ so they could fall back on the Kato theory.

We first remark that, the four R-classes do, in fact, meet our demand $V(r) = r^{-\alpha} \in R + (L^\infty)_\varepsilon$ for all $\alpha < 2$. We also mention that these classes are (with one exception) extensions of their Kato class counterparts as the following diagram indicates:

$$\begin{array}{ccc}
L^2 + L^\infty & \subset & R + L^\infty \\
\cup & & \cup \\
L^2 + (L^\infty)_\varepsilon & \subset & R + (L^\infty)_\varepsilon \\
\cup & & \cup \\
L^2 & & R \\
\cup & & \cup \\
L^1 \cap L^2 & \subset & R \cap L^1 .
\end{array}$$

To help understand this diagram, we note that as one goes up, worse behavior at infinity is allowed and as one goes to the right, worse behavior at finite points is allowed.

In Chapter I, we make a detailed comparison of this Rollnik condition with various other conditions which have been considered by other authors. We would urge the reader to avoid becoming bogged down in the technical details of this chapter; he might well look at the examples in §I.6 turning back to the theorems justifying the remarks when he desires.

In a sense the core of the monograph is in Chapter II where we discuss the definition of the Hamiltonian. Instead of defining $H_o + V$ as the operator sum on $D(H_o) \cap D(V)$, we show there is a unique self-adjoint operator H so that $\langle\psi, H\psi\rangle = \langle\psi, H_o \psi\rangle + \langle\psi, V\psi\rangle$ for all $\psi \epsilon D(H_o^{1/2})$; thus the Hamiltonian is the "sum" of the kinetic and potential energy in the natural sense for observables. One can ask how this quadratic form definition compares with other possible definitions of H. Anyone who has kept abreast with recent developments in constructive quantum field theory might suggest defining H by cutting off V and proving $H_o + V_n$ converges in a suitable sense as the cutoff goes away. It follows from results in §II.4 that not only do the resolvents of $H_o + V_n$ converge in norm but they converge to the resolvent of the operator defined as a sum of forms. In §II.6, we see that attempts to define $H_o + V$ as a Feynman path integral invariably lead one to the quadratic form definition of H.

For a summary of the material in Chapters III-VII we refer the reader to the introduction of each of those chapters.

We have been careful to write this monograph so that the reader who is interested in a particular subject can study the particular sections of interest to him if he is willing to thumb backwards occasionally. For example, someone interested in the proof of dispersion relations from first principles could proceed as follows: read §II.2 to obtain the definition of H and then turn to §IV.1 where we have attempted to give an exposition of time-dependent scattering theory. The connection between scattering and the Lippman-Schwinger equation can be found in §V.5

although it will be necessary to refer to §IV.5 and §V.4. Chapter VI then provides a detailed proof of dispersion relations.

As the reader must know, in any work of this size the author is indebted to a large group of colleagues. Two individuals played important roles in the early stages. It was G Tiktopoulos who first pointed out that $r^{-3/2}$ limit for Kato potentials was not physically natural, and who suggested the class $R \cap L^1$ as a candidate for a more general theory because one could derive forward dispersion relations for them. And E. Nelson taught me that it was perfectly honorable to define Hamiltonians as sums of quadratic forms.

In addition, I have profited from conversations with V. Bargmann, C. Fefferman, J. J. Kohn, M. Reed, and E. Stein, and from correspondence with T. Kato and T. Ikebe.

Finally, it is a pleasure to be able to acknowledge my enormous debt to A. S. Wightman. Throughout my attempt to understand the mathematics of non-relativistic quantum mechanics, he has been a source of constructive suggestions, new ideas and friendly encouragement. And his careful reading of the manuscript for this monograph has enriched the final product considerably.

CONTENTS

Quantum Mechanics for Hamiltonians Defined as Quadratic Forms

CHAPTER I

THE ROLLNIK CONDITION

I.1 *Relation to* L^p *spaces*

In this chapter, we will consider in detail various properties of measurable functions, V(x), obeying:

$$\int \frac{|V(x)|\ |V(y)|}{|x-y|^2}\ d^3x\ d^3y < \infty . \tag{I.1}$$

Such potentials were first considered by Rollnik [98] and so we call (I.1) the Rollnik condition. We denote the set of such potentials by R and define the Rollnik norm:

$$\|V\|_R^2 = \int \frac{|V(x)|\ |V(y)|}{|x-y|^2}\ d^3x\ d^3y . \tag{I.2}$$

In Section I.2, we will show that R with $\|\ \|_R$ is a complete normed vector space. We remark that the significance of (I.1) is that it assures us that $V_{\|}^{1/2}\ (E - H_0)^{-1} V^{1/2}$ is a Hilbert-Schmidt operator in the limit $E \uparrow 0$ (and for any other E!);[1] this aspect of the Rollnik condition will be considered in Section I.4.

In this section, we study the relation of the Rollnik potentials to L^p conditions. The L^p spaces, we recall, are defined by:

DEFINITION. $f \in L^p$ if and only if $\int |f(x)|^p\ d^3x < \infty$. In this case, we set $\|f\|_p = (\int |f(x)|^p dx)^{1/p} (1 \le p < \infty)$. L^∞ is defined as the set of functions bounded almost everywhere with $\|f\|_\infty = \inf \{L \mid |f(x)| < L$ almost everywhere$\}$.

[1] We adopt the following shorthand notation throughout this monograph:

$$V_{\|}^{1/2}(x) = |V(x)|^{1/2};\ V^{1/2}(x) = V_{\|}^{1/2}(x)\ [\mathrm{sgn}\ V(x)] = V(x)/V_{\|}^{1/2}(x).$$

The basic result relating L^p and R is:

THEOREM I.1 (Kato [70]). *If* $V \epsilon L^{3/2}$, *then* $V \epsilon R$. *In fact (for a constant* C *independent of* V):

$$\|V\|_R \leq C\|V\|_{3/2}. \tag{I.3}$$

Proof. This is a direct consequence of the Sobolev inequalities which we discuss in Appendix 2 of this section.∎

Remark.

In Section I.6, we will present an example of a V which is in R, but is (barely) not in $L^{3/2}$.

Given a function, f, we can define $f_>(x) = f(x)$ when $|f(x)| > 1$ and 0 otherwise. Also, let $f_< = f - f_>$. Then $f \epsilon L^p$ implies $f_> \epsilon L^q$ for $q \leq p$ and $f_< \epsilon L^r$ for $r \geq p$. Thus Theorem I.1 has two immediate consequences:

COROLLARY I.2. *If* $p \geq \frac{3}{2}$, $L^p + L^\infty \subset R + L^\infty$; *in particular*

$$L^2 + L^\infty \subset R + L^\infty.$$

Proof. Let $V = f + g$; $f \epsilon L^p$, $g \epsilon L^\infty$. Then $f_> \epsilon L^{3/2}$ and $f_< + g \epsilon L^\infty$.∎

Remark.

By putting an ϵ in the break-up $f_> + f_<$, we can show $L^2 + (L^\infty)_\varepsilon \subset R + (L^\infty)_\varepsilon$ where $X + (L^\infty)_\varepsilon = \{f | (\forall \varepsilon)\ f = f_1 + f_2$ with $f_1 \epsilon X$ and $\|f_2\|_\infty < \varepsilon\}$.

COROLLARY I.3. *If* $p \leq \frac{3}{2} \leq q$, *then* $L^p \cap L^q \subset R$; *in particular*

$$L^1 \cap L^2 \subset R \cap L^1.$$

Proof. If $V \epsilon L^p \cap L^q$, then $V_> \epsilon L^{3/2}$ (since $V \epsilon L^q$) and $V_< \epsilon L^{3/2}$ (by $V \epsilon L^p$) so $V \epsilon L^{3/2} \subset R$.∎

The crucial case of Corollary I.3, that $L^1 \cap L^2 \subset R$, has been noted by Grossman and Wu [45]. There is a useful modification of their proof which allows us to relate $\|\ \|_R$, $\|\ \|_2$, and $\|\ \|_1$ in a new way:

THEOREM I.4. *Let* $V \epsilon L^1 \cap L^2$. *Then:*

$$\|V\|_R \le 3^{1/2} (2\pi)^{1/3} (\|V\|_2)^{2/3} (\|V\|_1)^{1/3} . \tag{I.4}$$

Proof. Let r be arbitrary. Then:

$$\int_{|x-y|>r} \frac{|V(x)|\,|V(y)|}{|x-y|^2} d^3x\, d^3y \le \frac{1}{r^2} \left(\int |V(x)| d^3x\right)^2 = \frac{\|V\|_1^2}{r^2} .$$

On the other hand:

$$\int_{|x-y|\le r} \frac{|V(x)|^2}{|x-y|^2} d^3x\, d^3y = \|V\|_2^2 \int_{|z|\le r} \frac{d^3z}{|z|^2} = 4\pi r \, \|V\|_2^2 .$$

Thus, as a function on $\{<x, y>|\ |x-y| \le r\}$, $|V(x)| / |x-y| \epsilon L^2$ and so is $|V(y)| / |x-y|$. Cauchy-Schwartz then implies:

$$\int_{|x-y|\le r} \frac{|V(x)|\,|V(y)|}{|x-y|^2} d^3x\, d^3y \le 4\pi r \, \|V\|_2^2 .$$

As a result, for any r, we have:

$$\|V\|_R^2 \le r^{-2} \|V\|_1^2 + 4\pi r \, \|V\|_2^2 .$$

This is optimized by $r = (\|V\|_1^2 / 2\pi \|V\|_2^2)^{1/3}$ which yields (I.4).∎

Remark.

Theorem I.4 (with a different constant) can be proven alternately by using (I.3), for:

$$\int_{\{x|\ |V(x)|>k\}} |V(x)|^{3/2} dx \le k^{-1/2} \|V\|_2^2$$

$$\int_{\{x|\ |V(x)|\le k\}} |V(x)|^{3/2} dx \le k^{1/2} \|V\|_1 .$$

so that $\|V\|_{3/2}^{3/2} \le k^{-1/2} \|V\|_2^2 + k^{1/2} \|V\|_1 \le 2\|V\|_2 \|V\|_1^{1/2}$ (optimizing k).

Thus $\|V\|_{3/2} < C\, \|V\|_2^{2/3} \|V\|_1^{1/3}$ which with (I.3) implies a result of the same type as (I.4).

One can ask to what extent $V \epsilon R$ is weaker than $V \epsilon L^{3/2}$. Recall that:

DEFINITION. $f \epsilon (L^p)_W$ (weak L^p) if and only if $\mu\{x \mid |f(x)|>t\} < c/t^p$ for some c.

We discuss these spaces in Appendix 1 of this section.

THEOREM I.5. *Let* $V(x)$ *be central* [*i.e.* $V(x) = V(|x|)$] *and monotone* (*i.e.* $|V(r_1)| > |V(r_2)|$ *if* $r_1 < r_2$). *Then* $V \epsilon R$ *implies* $V \epsilon (L^{3/2})_W + L^\infty$.

Proof. Let $C = \int_{|x|<1;|y|<1} d^3x\, d^3y/|x-y|^2$ so

$$\int_{|x|<r;|y|<r} d^3x\, d^3y/|x-y|^2 = Cr^4.$$

If V is bounded it is in L^∞, so suppose $V \to \pm\infty$ at $r = 0$. Let r_t be defined by $|V(r_t)| = t$, so $r < r_t$ implies $|V(r)| > t$. Then $V \epsilon R$ implies:

$$\|V\|_R^2 \geq \int_{|x|<r_t;|y|<r_t} \frac{|V(x)|\,|V(y)|}{|x-y|^2}\, d^3x\, d^3y \geq (t^2)\, C(r_t)^4.$$

Thus $r_t \leq c'\, t^{-1/2}$ so $\mu\{x \mid |V(x)|>t\} = \frac{4}{3}\pi r_t^{\,3} < C''\, t^{-3/2}$, i.e. $V \epsilon (L^{3/2})_W$. ∎

Monotonicity and centralness are not really essential. What is really important is that V blows up without wiggling too much and without blowing up on "thin sets." Specifically, let us define:[2]

DEFINITION. A potential, V(x) will be called non-nasty if and only if there are constants t_o, λ, E so that, for every $t > t_o$, there is an r_t

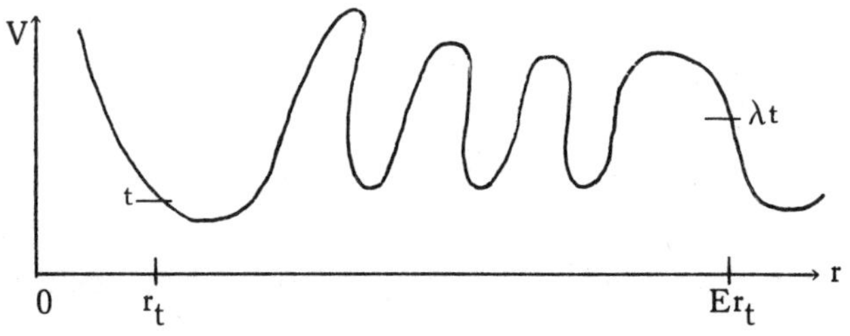

---

2 We could allow blow-ups at a finite number of points rather than just $\vec{r} = 0$.

with $|V(x)| > t$ for $|x|<r_t$ and $|V(x)|<\lambda t$ for $|x|> Er_t$. Thus, for each fixed t, we picture the schematic behavior shown on preceding page.

The proof of Theorem I.5 can be modified easily to yield:

THEOREM I.6. *Let* $V \epsilon R$ *and suppose* V *is non-nasty. Then* $V\epsilon (L^{3/2})_W + L^\infty$. *Thus* $\int_{C_0 \leq |V(x)| \leq \Lambda} |V(x)|^{3/2}\, dx$ *diverges at worst logarithmically as* $\Lambda \to \infty$. *If* $V \epsilon R \cap L^1$ *is non-nasty, then* $V \epsilon L^p$ *for any* $1 \leq p < 3/2$.∎

* * * *

One last (and rather trivial) relation to the L^p spaces will be needed:

THEOREM I.7. *Let* $V \epsilon R + L^\infty$ *and let* V *be of bounded support* (*i.e. for some* R, $|x| > R$ *implies* $V(x) = 0$). *Then* $V \epsilon R \cap L^1$. *In particular, any* $V \epsilon R + L^\infty$ *is locally* L^1.

Proof. Since any L^∞ function of compact support is in R (it is in $L^{3/2}$), we can suppose without loss that $V \epsilon R$. Since $\|V\|_R < \infty$, we know $\int dy\, |V(y)| / |x-y|^2 < \infty$ for some x (by Fubini's theorem). Thus, since $|x-y|^2 < (|x| + R)^2$ where $V(y) \neq 0$, $\int dy\, |V(y)| < \infty$, i.e. $V \epsilon L^1$.∎

Appendix 1 to I.1. The weak L^p *spaces*

Since most physicists are unfamiliar with the weak L^p spaces (but have a passing acquaintance with the usual L^p spaces), we present a brief discussion of their properties.

The distribution function, $m_f(t)$ (or just $m(t)$) for a measurable function f on a measure space (M,μ) is defined by:

$$m_f(t) = \mu\{x \mid |f(x)| > t\} \, . \tag{I.5}$$

An elementary computation proves:

LEMMA I.8. $f \in L^p(\mu)$ *if and only if* $\int_0^\infty t^{p-1}\, m(t)\, dt < \infty (1 \leq p < \infty)$ *and* $\|f\|_p^p = p \int_0^\infty t^{p-1}\, m(t)\, dt$.

Proof. $\int |f(x)|^p\, dx = \int_0^\infty t^p\, dm(t)$ (Stieltjes integral)

$$= p \int_0^\infty t^{p-1} m(t)\, dt \; .$$

(The boundary terms can be argued away.) ∎

The weak L^p spaces are defined by:

DEFINITION. $f \in (L^p)_W$ if and only if $m_f(t) \leq c/t^p$ for some $c < \infty$.

Thus $L^p \subset (L^p)_W$, but there are $f \in (L^p)_W$ with $\int_0^\infty t^{p-1}\, m(t)\, dt$ logarithmically divergent (at 0 and/or ∞) so that the inclusion is strict. For example $(L^p)_W\,(R)$ includes functions like $x^{-1/p}$. The real value of the weak spaces is that various inequalities for $f \in L^p$ extend to $f \in (L^p)_W$. Thus integrals that a priori might be logarithmically divergent turn out to converge.

There is a very simple characterization of $(L^p)_W$ contained implicitly in the work of Calderon [10], Lions and Peetre [85], and Stein and Weiss [105]:

LEMMA I.9. *Let* $p_0 < p < p_1$. *Then* $f \in (L^p)_W$ *if and only if, there exist* C_0, C_1 *so that, for all* λ, $f = f_{0,\lambda} + f_{1,\lambda}$ *with*

$$\|f_{i,\lambda}\|_{p_i} < C_i(\lambda)^{1-(p/p_i)} \quad i = 0, 1.$$

Proof. An elementary manipulation, see e.g. Nelson [T23]. ∎

The idea behind the proof is that, in the language of Section I.1, when $f \in (L^p)_W$, $f_> \in L^q$ for $q < p$ and $f_< \in L^q$ for $q > p$.

Appendix 2 to I.1. The Sobolev Inequality

Let us discuss the basic inequalities

$$\int \frac{|f(x)|\ |h(y)|}{|x-y|^{\lambda}}\ d^n x\ d^n y \le C_{p,r,\lambda,n}\ \|f\|_p\ \|h\|_r$$

for $f \in L^p(R^n)$, $h \in L^r(R^n)$ and

$$\frac{1}{p} + \frac{1}{r} + \frac{\lambda}{n} = 2;\ \lambda < n.$$

The earliest proof of this result is due to Hardy and Littlewood [48] in the $n = 1$ case (see also Hardy et al [T14], eq. 382, pg. 288) and to Sobolev [104] who reduced the general case to the $n = 1$ case. Du Plessis [95] later found a completely elementary trick to reduce any n to $n = 1$ (see Bers et al [T3], pp. 242-243). There is an elegant proof due to Stein and Weiss [105] which uses a strengthened Marcinkiewicz interpolation theorem. We present a simple proof[3] using two interpolation formulae:

Marcinkiewicz interpolation theorem (see Marcinkiewicz [88], Zygmund [130]). Let $p_i < q_i$, $i = 1, 2$. Let[4]

$$T: L^{p_i} \to (L^{q_i})_W,\ \text{bounded}^5\ i = 1, 2\,.$$

Let $q_1 \neq q_2$. Then for any $0 < t < 1$, $T: L^p \to L^q$ is a bounded map where

$$\frac{1}{p} = \frac{t}{p_1} + \frac{(1-t)}{p_2};\ \frac{1}{q} = \frac{t}{q_1} + \frac{(1-t)}{q_2}\,. \tag{I.5}$$

---

[3] I should like to thank Prof. E. M. Stein for a valuable discussion about the material below. A preliminary version used Fourier transforms and required $p, q \le 2$. Prof. Stein suggested that I use convolutions directly.

[4] To say we have the same map on different L^p spaces means that we have a map on finite sums of characteristic functions of sets of finite measure which we extend by continuity.

[5] To say $T: L^p \to (L^q)_W$ is bounded means that there is a C, independent of f, so that $m_{Tf}(\lambda) \le C\,[\|f\|_p/\lambda]^q$ for all $f \in L^p$.

Hunt's Interpolation Theorem (Hunt [50, 51], Calderon [10], Lions and Peetre [85], Stein and Weiss [105]).

Let $q_1 \neq q_2$, $p_1 \neq p_2$. Let:

$$T: L^{p_i} \to L^{q_i}, \text{ bounded, } i = 1, 2.$$

Then for $0 < t < 1$, $T: (L^p)_W \to (L^q)_W$ is bounded where p and q are given by (I.5).

For a general discussion of interpolation formulae, see Nelson [T23]; for the Marcinkiewicz theorem, also see Zygmund [T30], pp. 111-120. We call Hunt's theorem by that name (it is not standard) since it is a special case of a theorem of R. Hunt [50]. This case can be proven easily from Lemma I.9.

CONVENTION. For the remainder of this appendix, p and p′ (or q, q′, etc.) will stand for a conjugate pair of indices, i.e. $p^{-1} + (p')^{-1} = 1$.

LEMMA I.10. *Let* $f \in L^p$. *If* $g \in L^{p'}$, *then* $f * g \in L^\infty$, *and* $\|f*g\|_\infty \leq \|f\|_p \|g\|_{p'}$. *If* $f \in L^p$ *and* $g \in L^1$, *then* $f * g \in L^p$ *and* $\|f * g\|_p \leq \|f\|_p \|g\|_1$.

Proof. If $g \in L^{p'}$. by Hölder's inequality:

$$\left|\int f(x)\, g(y-x)\, dx\right| \leq \|f\|_p \, \|g\|_{p'}$$

which proves the first statement. Let $g \in L^1$ and $h \in L^{p'}$ be arbitrary. Then:[6]

$$\begin{aligned} \left|\int h(x)\, (f*g)\,(x)\right| &\leq \int |h(x)|\, |f(x-y)|\, |g(y)|\, dx\, dy \\ &\leq \int dy\, \left[\int |h(x)\, f(x-y)| dx\right] |g(y)| \\ &\leq \|g\|_1 \, \|\, |h|*|f|\, \|_\infty \leq \|g\|_1 \, \|h\|_{p'} \|f\|_p . \end{aligned}$$

Thus $f*g \in L^p$ and $\|f*g\|_p \leq \|g\|_1 \, \|f\|_p$. ∎

---

6 This is just a duality argument. $(f*): L^{p'} \to L^\infty$ so $(f*)^\dagger: L^1 \to L^p$, but $(\overline{f*})^\dagger = f*$.

Remark.

By a third interpolation theorem (the Reisz-Thorin theorem) which we haven't mentioned (or alternately by combining our two interpolation theorems), we can prove $L^p * L^q \subset L^s$ if $p^{-1} + q^{-1} > 1$; $p^{-1} + q^{-1} = 1 + s^{-1}$ using Theorem I.10. This result is Young's theorem.

LEMMA I.11. *Let* $f \epsilon L^p$. *Let* $g \epsilon (L^q)_W$ *where* $1 < q < p'$. *Then* $f*g \epsilon (L^s)_W$ *where* $p^{-1} + q^{-1} = 1 + s^{-1}$.

Proof. This follows from Hunt's interpolation theorem and Lemma I.10. ∎

LEMMA I.12. *Let* $f \epsilon L^p$. *Let* $g \epsilon (L^q)_W$ *where* $1 < q < p' < \infty$. *Then* $f*g \epsilon L^s$ *where* $p^{-1} + q^{-1} = 1 + s^{-1}$.

Proof. Fix $g \epsilon (L^q)_W$. Then $*g : L^p \to (L^s)_W$ where

$$s^{-1} = p^{-1} + (q^{-1}) < p^{-1}.$$

Thus $*g$ increases the index so Marcinkiewicz can be applied. Given p and q obeying the conditions, we find p_1, p_2 with $1 < p_1 < p < p_2 < q'$ and apply the Marcinkiewicz theorem using $*g : L^{p_i} \to (L^{s_i})_W$;

$$p_i < s_i, \ p_1 < p < p_2$$

to conclude the proof of the lemma. ∎

Remark.

Using Hunt's theorem, we could prove now a weak Young's theorem, i.e. $(L^p)_W * (L^q)_W \subset (L^s)_W$.

Lemma I.12 is equivalent to:

THEOREM I.13. *Let* $f \epsilon L^p$, $g \epsilon (L^q)_W$, $h \epsilon L^r$ *with* $p^{-1} + q^{-1} + r^{-1} = 2$; $1 < p,q,r < \infty$.

Then

$$|\int f(x)\ g(x-y)\ h(y)| < C_g \|f\|_p\ \|h\|_r \ .$$

Proof. $p^{-1} + q^{-1} > 1$ so $1 < q < p' < \infty$. Thus $f*g \epsilon L^{r'}$. ∎

COROLLARY I.14. *Let* $p > 1$, $r > 1$, $\lambda < n$, $p^{-1} + r^{-1} + (\lambda/n) = 2$. *Let* $f \epsilon L^p(R^n)$, $h \epsilon L^r(R^n)$.
Then:

$$\int \frac{|f(x)|\ |h(y)|}{|x-y|^\lambda} d^nx\ d^ny \leq C_{p,r,\lambda,n} \|f\|_p\ \|h\|_r \ .$$

Proof. Let $g(z) = 1/|z|^\lambda$. Then, $m_g(t)$ = volume of a sphere of radius $t^{-1/\lambda} = C_n\ t^{-n/\lambda}$. Thus $g \epsilon (L^{n/\lambda})_W$ and Theorem I.13 implies the result. ∎

I.2. *The p-space form*

In this section, we will examine the various p-space forms of the Rollnik condition. These connections between the p-space and x-space forms and, in particular, the fact that $\int \frac{V(x)\ V(y)}{|x-y|^2} d^3x\ d^3y \geq 0$ appear to have been first noted explicitly in the physics literature by Ghirardi and Rimini [42].[7] Our treatment differs from theirs mainly in its rigor (and thereby in its pedantry).

LEMMA I.14. *Let* $V \epsilon L^1 \cap L^2$. *Let*[8]

$$\hat{V}(p) = \int V(x)\ e^{-ip\cdot x}\ d^3x \ .$$

Then for any $a > 0$:

$$\int \frac{V(x)\ V(y)}{a^2 + |x-y|^2} d^3x\ d^3y = \frac{1}{4\pi} \int \frac{|\hat{V}(p)|^2}{p} e^{-ap}\ d^3p \ . \tag{I.6}$$

---

7 One dimensional analogues of these formulae appear often in the mathematical literature, see e.g. Kunze and Stein [77].

8 This definition of Fourier transform differs from the one we will use in Chapter IV by a factor of $(2\pi)^{-3/2}$. We will also use ^ for a different operation at that time.

Proof.[9] With our normalization of $\hat{}$, we have for any $f,\ h\in L^2,\ g\in L^1$:

$$\begin{aligned}\int dx\, dy\, \overline{f(x)}\, h(x-y)\, g(y) &= \int dx\, \overline{f(x)}\, (h*g)(x)\\ &= \frac{1}{(2\pi)^3}\int dp\, \overline{\hat{f}(p)}\, \widehat{h*g}(p)\\ &= \frac{1}{(2\pi)^3}\int dp\, \overline{\hat{f}(p)}\, \hat{h}(p)\, \hat{g}(p)\,.\end{aligned}$$

Letting $f = g = V$ and $h(u) = (|u|^2+a^2)^{-1}$, we obtain the result by computing (all integrals in $\lim_{R\to\infty}\int_{|x|<R}$ sense):

$$\hat{h}(p) = \int d^3u\, \frac{e^{-ip\cdot u}}{|u|^2+a^2} = \frac{4\pi}{|p|}\int_0^\infty \frac{u\,du}{u^2+a^2}\sin(pu) = \frac{2\pi^2}{|p|}\, e^{-pa}.\ \blacksquare$$ [10]

THEOREM I.15. *Let* $V \geq 0,\ V\in L^1$. *Then* $V\in R$ *if and only if*

$$\int \frac{|\hat{V}(p)|^2}{p}\, d^3p < \infty. \tag{I.7}$$

Whenever $V\in R \cap L^1$ *(even if* V *is not positive):*

$$\int \frac{V(x)\, V(y)}{|x-y|^2}\, d^3x\, d^3y = \frac{1}{4\pi}\int \frac{|\hat{V}(p)|^2}{p}\, d^3p\,.\ \blacksquare \tag{I.8}$$

Proof. We first note that (I.6) holds whenever $V\in L^1$ by a simple limiting argument. When $V \geq 0$, both integrands have monotone limits as $a \downarrow 0$ and thus both are finite simultaneously, which proves the first statement. If $V\in R$, the integrand on the left hand side of (I.6) is dominated by $|V(x)|\ |V(y)|/|x-y|^2$ and so the limits still may be taken.

---

[9] I should like to thank Dr. C. Fefferman for a valuable discussion on the best way of approaching the p-space formulae.

[10] By Gradshteyn-Ryzhik [T10], pg. 406, §3.723 or a simple contour integral.

Remarks.

1. There presumably exist misbehaved $V \in L^1$ which are not positive so that (I-7) holds but for which the left hand side of (I-8) is not absolutely integrable.

2. The above can also be used to show that for any $V \in R$, $\hat{V}(p)$ exists in the sense that $\int |\hat{V}_n(p) - \hat{V}_m(p)|^2 p^{-1} d^3p \to 0$ for any reasonably cutoff that $\hat{V}_n(p)$ is locally L^2 convergent to a function for which (I.8) holds.

LEMMA I.16. *Let* $V \in R$. *Then*

$$\int \frac{V(x)\ V(y)}{|x-y|^2}\, d^3x\ d^3y \geq 0\ . \tag{I.9}$$

Proof. Let V be cutoff to yield a sequence V_n with $V_n(x) \to V(x)$ (pointwise), $|V_n(x)| < |V(x)|$ and $V_n(x) \in L^1$. Then

$$\int \frac{V(x)\ V(y)}{|x-y|^2}\, d^3x\ d^3y = \lim_{n \to \infty} \int \frac{|\hat{V}_n(p)|^2}{p}\, d^3p \geq 0. \blacksquare$$

THEOREM I.17. R *is a vector space and* $\| \ \|_R$ *is a norm on it.*

Proof. First suppose $V, W \in L^1 \cap L^2$ so that $V + aW \in R$ for any real a. Thus

$$\int \frac{[(V+aW)(x)]\ [(V+aW)(y)]}{|x-y|^2}\, d^3x\ d^3y \geq 0$$

any a, so that:

$$\left| \int \frac{V(x)\,W(y) + W(x)\,V(y)}{|x-y|^2}\, d^3x\ d^3y \right| \leq \|V\|_R\ \|W\|_R\ . \tag{I.10}$$

By a simple limiting argument, (I.10) holds for any $V, W \geq 0$. Thus if $V, W \in R$, $\int \frac{d^3x\ d^3y}{|x-y|^2} |V(x) + W(x)|\ |V(y) + W(y)|$ is finite and bounded by $(\|V\|_R + \|W\|_R)^2$. $\blacksquare$

Remark.

Notice that $\| \ \|_R$ is *not* realized as a Hilbert space norm; it is $\int \frac{V(x)\,V(y)}{|x-y|^2}\, d^3x\, d^3y$ (where V rather than $|V|$ appears) that is the Hilbert space norm.

THEOREM I.18. R *is complete in* $\| \ \|_R$.

Proof. One need only mimic the standard Riesz-Fischer proof for the completeness of L^2 (c.f. Berberian [T2] pp. 110-113). Given $V_n \epsilon R$ with $\|V_n - V_m\|_R \to 0$, we need only find a convergent subsequence. Pick a subsequence, still denoted by V_n with $\|V_n - V_{n+1}\|_R < 2^{-n}$. Then $W(x) = \sum_{n=1}^{\infty} |V_{n+1}(x) - V_n(x)|$ is the monotone limit of $\sum_{n=1}^{N} |V_{n+1} - V_n|$ so by the monotone convergence theorem

$$\int \frac{W(x)\,W(y)}{|x-y|^2}\, d^3x\, d^3y$$

$$= \lim_{N\to\infty} \Big\| \sum_{n=1}^{N} |V_n - V_{n+1}| \Big\|_R^2 \le \lim_{N\to\infty} \Big(\sum_{n=1}^{N} \|V_n - V_{n+1}\|_R\Big)^2 < \infty .$$

Thus $W \epsilon R$ and so $W(x) < \infty$ a.e. Thus $V_1(x) + \sum_{n=1}^{\infty} V_{n+1}(x) - V_n(x)$ is absolutely convergent a.e. to a sum $V(x)$ with $|(V - V_1)(x)| < W(x)$. Thus $V \epsilon R$ and it is not hard to show $V_n \to V$ in $\| \ \|_R$. ∎

I.3. *Relation to the Born series*

In a brief paper, Scadron, Weinberg and Wright [100] study Rollnik's condition (not by that name!) and its relation to the compactness of $V_{\|}^{1/2} (E + H_0)^{-1} V^{1/2}$ (see Section I.4). They note that the Born series is absolutely convergent for small coupling constants when $V \epsilon L^1 \cap R$. In this section we restate their proof and consider the extent to which the converse is true. For non-nasty potentials, we will precisely characterize the potentials with convergent Born series. We first recall that the Born series for energy $E = k^2$ (with $\mathrm{Im}\ k \geq 0$) is given by

$$B(E,\lambda) = \sum_{n=1}^{\infty} B_n(E)\,\lambda^n \tag{I.11}$$

where

$$B_n(E) = (-1)^n \int d^3x_1 \ldots d^3x_n\ V(x_1)\frac{e^{ik|x_1-x_2|}}{4\pi\,|x_1-x_2|}V(x_2)$$

$$\ldots\frac{e^{ik|x_{n-1}-x_n|}}{4\pi\,|x_{n-1}-x_n|}\ V(x_n)\ . \tag{I.12}$$

DEFINITION. We will say V obeys condition (CB) if and only if:

(1) $|B|_n = \int d^3x_1 \ldots d^3x_n \dfrac{|V(x_1)|\ldots|V(x_n)|}{|x_1-x_2|\ldots|x_{n-1}-x_n|} < \infty$.

(2) For some Λ, $\Sigma\lambda^n|B|_n < \infty$ for all $|\lambda| < \Lambda$.

Then Scadron, et. al prove:

THEOREM I.19. *Let* $V \epsilon L^1 \cap R$. *Then* (CB) *is obeyed* (*with*

$$\Lambda = (4\pi)/\|V\|_R).$$

Thus, for any E, *each of the integrals* $B_n(E)$ *is absolutely convergent and when* $\lambda\|V\|_R < (4\pi)$, $B(E,\lambda)$ *is absolutely convergent.*

Proof. Without proof, we recall two elementary facts:

(a) If $K_i(x,y)$, $i = 1, 2$ are square integrable in x and y (jointly) then so is $M(x,y) = \int K_1(x,z)\,K_2(z,y)\,dz$ and $\|M\|_{sch} \leq \|K_1\|_{sch}\|K_2\|_{sch}$ (where $\|K\|_{sch}^2 \equiv \int\|K(x,y)\|^2\,dx\,dy$). In this case we write $M = K_1 \circ K_2$.

(b) If $K(x,y)$ is square integrable and $\phi, \psi \epsilon L^2$, then

$$|\int dx\,dy\,\phi(x)\ K(x,y)\psi\,(y)| \leq \|\phi\|_2\ \|K\|_{sch}\|\psi\|_2.$$

If $V \epsilon R$, let $K_1(x,y) = |V(x)|^{1/2}\ |x-y|^{-1}\ |V(y)|^{1/2}$. Then $\|K_1\|_{sch} = \|V\|_R$

Let $K_n = K_1 \circ \cdots \circ K_1$ (n-times). Then $\|K_n\|_{sch} \le \|V\|_R^n$ by iterating (a). Thus, if $V \epsilon L^1$ also, $|B|_n =$

$$\frac{1}{(4\pi)^n} \int dx_1 \, dx_n \, |V(x_1)|^{\frac{1}{2}} K_n(x_1,x_n) \, |V(x_n)|^{\frac{1}{2}} \le \|V\|_1 \left[\frac{\|V\|_R}{4\pi}\right]^n .$$

Thus, the $|B|_n$ integrals are convergent and $\Sigma\lambda|B|_n < \infty$ if

$$\lambda\|V\|_R (4\pi)^{-1} < 1. \blacksquare$$

The converse is false as the following example shows:

Example. Let $V(r) = r^{-2}$ for $r < 1$ and 0 otherwise. Then:

(a) $V \not\epsilon R$; this will follow from general statements made in Section II.1 when we return to this example. It is also a result of the fact that its Fourier transform $\sim p^{-1}$ for p large[11] so

$$\int \frac{|\hat{V}(p)|^2}{p} d^3p = \infty, \text{ i.e. } V \epsilon R.$$

(b) V obeys (CB). This will follow from general considerations below. Alternately, let us show how the Sobolev inequalities at least imply the $|B|_n$ are finite. For suppose $f \epsilon L^p$, $p < 3/2$. Then Sobolev tells us

$f_1(x_{n-1}) \equiv \int dx_n \, f(x_n)/|x_{n-1}-x_n|$ has $f_1 \epsilon L^{q_1}$ where $q_1^{-1} = p^{-1} - 2/3$.

Then $ff_1 \epsilon L^{p_2}$ with $p_2^{-1} = q_1^{-1} + p^{-1}$. By Sobolev,

$f_2(x_{n-2}) = \int dx_n \, dx_{n-1} \, f(x_n) \, f(x_{n-1})/|x_n - x_{n-1}| \, |x_{n-2} - x_{n-2}| =$

$\int dx_{n-1} \, f(x_{n-1}) \, f_1(x_{n-1})/|x_{n-1} - x_{n-2}| \, \epsilon L^{q_2}$ with $q_2^{-1} = p_2^{-1} - 2/3 =$ $2[p^{-1} - 2/3]$. Thus $|B|_n$ is absolutely convergent if

[11] Up to a multiplicative constant,

$$\hat{V}(p) = p^{-1} \int_0^1 (r^{-1}) \sin pr \, dr = p^{-1} \int_0^p x^{-1} \sin x \, dx .$$

This last integral is well-known to converge to a constant as $p \to \infty$.

$(n-1)(p^{-1}-2/3)+p^{-1}=1$, and $V \in L^p$. Since the V under consideration is in L^p for any $p < 3/2$, all the $|B|_n$ converge. To show a finite radius of convergence, we would need to make a detailed study of constants in the Sobolev inequalities and of the L^p norms of V.

This example shows that the class of $V \in R \cap L^1$ and the class of V's with convergent Born series aren't precisely the same. However, they aren't very different; in fact:

THEOREM I.20. *Let* V(x) *be non-nasty. Then the following are equivalent:*

(1) V *obeys* (CB).

(2) $V \in L^1 \cap (L^{3/2})_W$.

(3) $V \in L^1$ *and* $|V(x)| < d + c|x|^{-2}$ *for some* c, d.

(4) $D(|V|^{1/2}) \supset D(H_0^{1/2})$ *and for some* a, b *and all* $\psi \in D(H_0^{1/2})$

$$\| |V|^{1/2}\psi\| \le a\|H_0^{1/2}\psi\| + b\|\psi\| \tag{I.13}$$

and $V \in L^1$ (i.e. $V \in L^1$ *and* $\lambda|V|^{1/2} \underset{\text{T.K.}}{<} |H_0|^{1/2}$ *for some* λ)

Proof. We will suppose V is monotone and central – the modifications for non-nasty V are easy. We prove the string of implications (1) → (2) → (3) → (4) → (1) using the r_t notation of Theorem I.5. [i.e. $r_t \equiv$ radius at which $V(r) = t$] (1) → (2). Since $|x|, |y| < R$ imply $|x-y|^{-1} > (2R)^{-1}$, we have

$$\int_{|x_j|<R} \frac{d^3x_1 \ldots d^3x_n}{|x_1-x_2|\ldots|x_{n-1}-x_n|} \ge 2^{n-1}R^{2n+1}(4\pi/3)^n.$$

Thus $2|B|_n > t^n(r_t)^{2n+1}(2\pi/3)^n$, so $r_t < (3/2\pi)^{n/2n+1}\, t^{-n/2n+1}(2|B|_n)^{1/2n+1}$. Since $\Sigma\, \lambda^n |B|_n$ has a non-zero radius of convergence, $\sup |B|_n^{1/2n+1} < \infty$, so $r_t < Ct^{-1/2}$. As a result, $V_> \in (L^{3/2})_W$. Since $|B|_1 < \infty$, $V \in L^1$.

In particular, then, $V_< \epsilon L^1$ and so $V \epsilon (L^{3/2})_W$. (2) → (3) $V \epsilon (L^{3/2})_W$ implies $r_t \leq ct^{-\frac{1}{2}}$ for some c. Thus for any r, $r \geq r_t$ with $t = (c/r)^2$ or $|V(r)| \leq c^2/r^2$. In the case where V is not monotone at large r (which is allowed with non-nastiness), this argument only shows that $|V(r)| < c^2/r^2$ at small distances. (3) → (4). It is clearly sufficient to prove an inequality of form (I-13) for $V = r^{-2}$. Such an inequality is proven by Courant and Hilbert [T6], pp. 446-447 (and is a quantitative form of the uncertainty principle). (4) → (1). Let E be positive. Then Ran $((E+H_o)^{-\frac{1}{2}}) = D(H_o^{\frac{1}{2}})$ so (I1.13) implies, for all ϕ:

$$\| \, |V|^{\frac{1}{2}} (E+H_o)^{-\frac{1}{2}} \phi \, \| < a \|H_o^{\frac{1}{2}} (E+H_o)^{-\frac{1}{2}} \phi\| +$$

$$b\| (E+H_o)^{-\frac{1}{2}} \phi\| \leq (a + bE^{-\frac{1}{2}})\|\phi\| \, .$$

As a result, $|V|^{\frac{1}{2}}(E + H_o)^{-\frac{1}{2}}$ is a bounded operator and thus so is $(E + H_o)^{-\frac{1}{2}} |V|^{\frac{1}{2}}$ (by taking adjoints). Let $V_1(x) = V(x)$ if $|x| < 1$ and 0 otherwise and let $V_2 = V - V_1$. We wish to show $|V(x)|^{\frac{1}{2}}|x-y|^{-1} |V(y)|^{\frac{1}{2}}$ is the kernel of a bounded operator. For this purpose, we need only show $|V_i(x)|^{\frac{1}{2}} |x - y|^{-1} |V_j(y)|^{\frac{1}{2}}$ has this property for i, j = 1, 2. The i = j = 2 term presents no problem since $V_2 \epsilon L^1 \cap L^\infty \subset R$. The i = j = 1 term can be handled as follows: since V_1 obeys (I.13), for any $\kappa > 0$,

$$|V_1(x)|^{\frac{1}{2}} e^{-\kappa|x-y|} |x-y|^{-1} |V_1(y)|^{\frac{1}{2}}$$

is the kernel of a bounded operator (by the above analysis). Since the kernel vanishes if $|x - y| > 2$, the $e^{-\kappa|x-y|}$ factor is minorized by $e^{-2\kappa}$; thus the i = j = 1 term yields a bounded operator. Finally, we must consider the cross terms. Let $V_2 = V_3 + V_4$, where $V_4(x) = V(x)$ if $|x| > 2$ and 0 otherwise. The $V_1 - V_3$ term can be treated in a manner similar to the $i = j = 1$ term. The $V_1 - V_4$ term defines a Hilbert-Schmidt kernel since $V_1, V_4 \epsilon L^1$ and $|V_1(x)|^{1/2} |x-y|^{-1} |V_4(y)|^{1/2} < |V_1(x)|^{1/2} |V_4(y)|^{1/2}$. Thus $A \equiv V_{\|}^{1/2} |x-y|^{-1} V_{\|}^{1/2}$ defines a bounded operator with bound C. $V \epsilon L^1$ implies $V_{\|}^{1/2} \epsilon L^2$. Since $|B|_n = \langle V_{\|}^{1/2}, A^{n-1} V_{\|}^{1/2} \rangle$, we have $|B|_n$ finite and $|B|_n \leq C^n \|V\|_1$ so (CB) holds. ∎

Remarks.

1. The (1) → (2) proof shows that if V is non-nasty and we only have $|B|_n < \infty$ for all n, then $V \epsilon L^q$ for all $1 \leq q < 3/2$.

2. Independently of V being non-nasty, the (4) → (1) proof shows that (4) yields finiteness of $B_n(E)$ for E unequal to a non-negative real and convergence of $B_n(E, \lambda)$ for $|\lambda|$ small (and E fixed).

3. We will see in Section II.1, for V non-nasty, that (4) is actually equivalent to $V \epsilon L^1$ and $D(|V|^{1/2}) \supset D(H_0^{1/2})$.

4. This theorem tells us that when a non-nasty potential obeys (CB), the Rollnik integral is at worst logarithmically divergent. Alternately, if V obeys (CB) and is non-nasty then $(1 + |\log r|)^{-1} V(r) \epsilon R$.

I.4. *Integral Kernels*

As we have already remarked, the real significance of the Rollnik condition is that it assures us that $|V(x)|^{1/2} |x-y|^{-1} e^{i\kappa|x-y|} |V(y)|^{1/2}$ and its variants are Hilbert-Schmidt kernels. This is the property that singles out R from the various classes of potentials differing from it by logarithms and this is the reason behind its frequent appearance in the literature; e.g. Rollnik [98], Schwinger [101], Grossman-Wu [44, 45] and Scadron et al. [100] all seemed to have arrived at this condition independently by studying this kernel. It is our goal to turn now to the study of the operators associated with this kernel and to investigate some general properties of the kernel (and its iterates) in detail.

THEOREM I.21. *Let* $V \in R$. *Then* $|V|^{1/2}$ *is Kato tiny relative to* $H_0^{1/2}$ [12]

Proof. Since $V \in R$, it is locally L^1 (Theorem I.7) and thus any $\psi \in C_0^\infty$ is in $D(|V|^{1/2})$. Moreover: $[(\kappa^2 + H_0)^{-1} |V|^{1/2}\psi](x) =$

$$(4\pi)^{-1} \int |x-y|^{-1} e^{-\kappa|x-y|} |V^{1/2}(y)| \psi(y)\, dy.$$

Thus, by the Rollnik condition $(\kappa^2 + H_0)^{-1} |V|^{1/2} \psi \in D(|V|^{1/2})$ and

$$<\psi, |V|^{1/2}(\kappa^2 + H_0)^{-1} |V|^{1/2} \psi >$$

$$\leq \frac{\|\psi\|^2}{4\pi} \left[\int \frac{|V(x)|\, e^{-\kappa|x-y|}\, |V(y)|}{|x-y|^2} d^3x\, d^3y \right]^{1/2}. \tag{I.13}$$

The integrand in (I.13) converges pointwise and monotonically to 0 as $\kappa \to \infty$. Thus, given a, we find κ so the left hand side of (I.13) is bounded by $a^2\|\psi\|^2$ (independent of ψ). Thus

$$\|(\kappa^2 + H_0)^{1/2} |V|^{1/2}\psi\| \leq a\|\psi\|$$

so that $(\kappa^2 + H_0)^{-1/2} |V|^{1/2}$ and its adjoint $|V|^{1/2} (\kappa^2 + H_0)^{-1/2}$ have continuous extensions with bound a. Thus, if $\psi \in D(H_0^{1/2})$:

$$\| |V|^{1/2}\psi\|^2 \leq \| |V|^{1/2}(\kappa^2 + H_0)^{-1/2} (\kappa^2 + H_0)^{1/2}\psi\|^2$$

$$\leq a^2 \|(\kappa^2 + H_0)^{1/2}\psi\|^2$$

$$\leq a^2 (\|H_0^{1/2}\psi\|^2 + \kappa^2 \|\psi\|^2)$$

$$\| |V|^{1/2}\psi\| \leq a\|H_0^{1/2}\psi\| + a\kappa\|\psi\| \tag{I.14}$$

which proves the result. ∎

---

12 See Appendix: Definition A.14.

Remarks.

1. An equivalent proof[13] of this result can be obtained by going to momentum space. R has another p-space form obtainable by first proving the result for $V \epsilon L^1 \cap L^\infty$ and then passing to the limit; viz. when $V \epsilon R$,

$$A_\kappa = \int \frac{|\hat{V}(p-q)|^2}{(p^2+\kappa^2)\,(q^2+\kappa^2)}\, d^3p\, d^3q < \infty \tag{I.15}$$

for all $\kappa > 0$ and (by the monotone convergence theorem) $A_\kappa \to 0$ as $\kappa \to \infty$. Then for $\psi \epsilon D(|p|)$ $(\equiv \widehat{D(H_0{}^{1/2})})$:

$$<\psi, |\hat{V}|\, \psi> = (2\pi)^3 \int \overline{\psi(p)}\, \psi(q)\, \hat{V}_{\|}\,(p-q)\, d^3p\, d^3q$$

$$= \int \frac{(2\pi)^3 \hat{V}_{\|}\,(p-q)}{(p^2+\kappa^2)^{1/2}\,(q^2+\kappa^2)^{1/2}} \left(\overline{\sqrt{p^2+\kappa^2}\psi(p)}\right) \left(\overline{\sqrt{q^2+\kappa^2}}\psi(q)\right) d^3p\, d^3q$$

$$\leq A_\kappa\, \|\sqrt{p^2+\kappa^2}\,\psi\|^2 .$$

2. (I.14) can be replaced with

$$\|\,|V|^{1/2}\psi\|^2 \leq a^2\, \|H_0{}^{1/2}\psi\|^2 + b^2\, \|\psi\|^2 \tag{I.16}$$

which is the form we will need.

THEOREM I.22. $V_{\|}{}^{1/2}(E-H_0)^{-1}\, V^1$ *and* $(H_0 - E)^{-1/2}\, V_{\|}\, (H_0 - E)^{-1/2}$ *define Hilbert-Schmidt operators (Definition A.28) for all* E *off the positive real axis. These functions are analytic in* E *and the first has a continuous extension to the real axis from above and below.*

Remarks.

1. The two extensions to the real axis are *not* the same, i.e. the operator valued function has a cut.

---

[13] Essentially a word for word translation of the x-space proof except that the technical details are transferred from Hilbert space to Fourier transforms.

2. $(H_0 - E)^{-1/2}$ is defined as multiplication by $(p^2 - E)^{-1/2}$ in p-space where the branch of square root is taken which is positive when $E < 0$.

Proof. The first operator has a kernel $V_{\|}^{1/2}(x)\, e^{i\kappa|x-y|}\, |x-y|^{-1}\, V^{1/2}(y)$ where $\kappa^2 = E$, $\operatorname{Im}\kappa \geq 0$. The results for it are thus clear. To study the second operator, we pass to p-space where the kernel is

$$(p^2 - E)^{-1/2}\, \hat{V}(p-q)\, (q^2 - E)^{-1/2}.$$

(I.15) implies that this is Hilbert-Schmidt. ∎

DEFINITION. Given $V \in R$, define for $\operatorname{Im}\kappa \geq 0$:

$$R_\kappa(x, y) = V_{\|}^{1/2}(x)\, \frac{e^{i\kappa|x-y|}}{|x-y|}\, V^{1/2}(y)\,. \tag{I.17}$$

There is an additional result about R_κ which we will need to know. It is the simple fact that $\lim_{\kappa \to \infty\, \kappa \text{ real}} \|R_\kappa\| = 0.$ Results of this form were first obtained by Zemach and Klein [129] (see also Aaron and Klein [1]) in their study of high energy convergence of the Born series. Since $\operatorname{Tr}(R^\dagger_\kappa R_\kappa)$ contains no κ-dependence, we can't use it to show $\|R_\kappa\| \to 0$. But

$$\operatorname{Tr}(R^\dagger_\kappa R_\kappa R^\dagger_\kappa R_\kappa) = \int d^3x\, d^3y\, d^3z\, d^3w\, \frac{e^{-i\kappa(|x-y| - |y-z| + |z-w| - |w-x|)}}{|x-y|\, |y-z|\, |z-w|\, |w-x|}$$

$$|V(x)|\, |V(y)|\, |V(z)|\, |V(w)|\,. \tag{I.18}$$

As $\kappa \to \infty$, we expect this integral to oscillate to death and, in fact, it does:

THEOREM I.23. (Zemach-Klein).[14] *Let* $V \in R$. *Then* $\lim_{\kappa \to \infty (\kappa \text{ real})} \|R_\kappa\| = 0.$

---

[14] This is not quite the Zemach-Klein result but the idea of the proof and the significance of the result are the same.

Proof. $V \in R$ implies the integrand in (I.14) is L^1 (since

$$(|V(x)|^{1/2}\,|x-y|^{-1}\,|V(y)|^{1/2}\,|V(z)|^{1/2}\,|z-w|^{-1}\,|V(w)|^{1/2})$$

and $(|V(z)|^{1/2}\,|x-z|^{-1}\,|V(x)|^{1/2}\,|V(y)|^{1/2}\,|y-w|^{-1}\,|V(w)|^{1/2})$ are both L^2). Make a change of variables replacing say $|x|$ by

$$R = |x-y| - |y-z| + |z-w| - |w-x|.$$

Then the integrand in (I.18) is equal to $\int_{-\infty}^{\infty} dR\, e^{i\kappa R} f(R)$ where $f(R)$ is the integral over all the other variables times a suitable Jacobian factor. Since the original integrand is L^1, $f \in L^1$ and thus by the Riemann-Lebesque lemma,[15] the integral goes to 0. Thus

$$\|R_\kappa\|^4 \equiv \|R^\dagger_\kappa R_\kappa\|^2 \leq \operatorname{Tr}(R^\dagger_\kappa R_\kappa R^\dagger_\kappa R_\kappa) \to 0. \blacksquare$$

I.5. *Potentials of Finite Range*

We will have some reason in Chapters IV and VI to study potentials which fall off exponentially at infinity.

DEFINITION. We say $V \in R$ has finite range if

$$\int |V(x)|\,|x-y|^{-2}\,|V(y)|\, e^{C|x|} e^{C|y|} < \infty$$

for some $C > 0$. We call the supremum over all C with this property the inverse range of V.

Such potentials have been extensively discussed by Grossman and Wu [44]. Most of the results below are due to them.

THEOREM I.24. *If* $V \in R$ *has inverse range* C, *then* $\int |V(x)|\, e^{D|x|}\, d^3x < \infty$ *for all* $D < C$.

---

[15] This says that $\lim_{\kappa\to\infty} \hat{f}(\kappa) = 0$ and follows from the L^1-mean convergence theorem, for $2f(\kappa) = \int \hat{f}(x)\,[e^{i\kappa x} - e^{i\kappa(x-\pi/\kappa)}] = \int e^{i\kappa x}\,[f(x) - f(x+\pi\kappa^{-1})]\,dx \to 0$.

Proof (Grossman-Wu). For any ε, $\kappa = \inf_{x,y} e^{\varepsilon|x|+\varepsilon|y|}\,|x-y|^{-2} > 0$.

$$\int \frac{|V(x)|\,e^{(D+\varepsilon)|x|}\,|V(y)|\,e^{(D+\varepsilon)|y|}}{|x-y|^2}\,d^3x\,d^3y \le \kappa\left(\int |V(x)| e^{D|x|}\,d^3x\right)^2 . \quad (I.19)$$

Since the left hand side of (I.19) is finite for some ε, the right hand side is also finite. ∎

THEOREM I.25. *Let* $V \epsilon R$ *and suppose:*

(a) *For some* D, $\int d^3x\ |V(x)|\ e^{D|x|} < \infty$.

(b) *For some* R_0 *and some* $p > 3/2$, $\int_{|x| > R_0} |V(x)|^p\ d^3x < \infty$.

Then V *has finite range.*

Proof. To say V has inverse range at least A means

$$W(x) = V(x)\ e^{A|x|} \varepsilon R.$$

Let $W = W_1 + W_2$ where W_1 vanishes for $|x| > R_0$ and W_2 vanishes for $|x| \le R_0$. Then $V \epsilon R$ implies $W_1 \epsilon R$ for any A. To treat W_2, let $q = 2(p-1) > 1$ and let r be the conjugate index to q. Then

$$q\left(\frac{3}{2} - \frac{1}{r}\right) = q\left(\tfrac{1}{2} + \frac{1}{q}\right) = 1 + \frac{q}{2} = p$$

so $|V|^{\frac{3}{2} - \frac{1}{r}} \varepsilon L^q$, for $|x| > R_0$. On the other hand $e^{Dr^{-1}|x|}\ |V|^{r^{-1}}\ \varepsilon L^r$ by (a) so $|V|^{3/2}\ e^{(Dr^{-1})|x|} \varepsilon L^1$ in $|x| > R_0$. Thus, for this value of A, $W \epsilon R$, i.e. V has finite range. ∎

One of the crucial properties of potentials of finite range is the analyticity of R_κ on the real axis, i.e.:

THEOREM I.26 (Grossman-Wu). *Let* $V \epsilon R$ *have inverse range* $C > 0$. *Then* R_κ *can be analytically continued to a compact-operator valued function in the region* Im $\kappa > -\frac{1}{2}$ C. *In particular,* R_κ *is analytic in a neighborhood of the real axis.*

Proof. Extend R_κ by the same functional form (I.17). This has Hilbert-Schmidt norm:

$$\int \frac{|V(x)|\,|V(y)|}{|x-y|^2}\, e^{-2\mathrm{Im}\kappa|x-y|}\, d^3x\, d^3y \leq$$

$$\int \frac{|V(x)|\,|V(y)|}{|x-y|^2}\, e^{-2\mathrm{Im}\kappa|x|}\, e^{-2\mathrm{Im}\kappa|y|}\, d^3x\, d^3y$$

when Im $\kappa \leq 0$. Thus, if 2Im $\kappa > -$ C, this is finite. The analyticity follows immediately (for example from the analyticity of matrix elements between functions of compact support and Theorem A.29). ∎

I.6. *Examples.*

We now present diverse examples which exhibit various facets of potentials obeying the Rollnik condition.

A. *Singular Potentials*

As we have remarked in the Introduction, a prime reason for considering Rollnik potentials is that physically non-singular potentials $V \sim r^{-\alpha}$; $2 > \alpha \geq \frac{3}{2}$ are singular (at $r = 0$) with respect to the Kato condition $V \epsilon L^2 + L^\infty$. We thus note the simple facts:

Example 1: If $V = Cr^{-\alpha}$, $\alpha < 2$, then $V \epsilon L^{3/2} + L^\infty$ and thus in $R + L^\infty$.[16]

Example 2: If $V_\beta(r) = r^{-2}(-\log r)^{-\beta}$; $(r < e^{-1})$, then $V_\beta \epsilon L^{3/2}$ if and only if $\beta > \frac{2}{3}$. In particular $V_\beta \epsilon R$ for $\beta > \frac{2}{3}$.

---

[16] Similarly at ∞, the condition for $V \epsilon R$ is $V \sim r^{-2-\varepsilon}$. Both Rollnik [98] and Weinberg [118] state stronger conditions at infinity than are needed.

This last example uses the computation:

$$\int |V_\beta|^{3/2}\, d^3r = 4\pi \int_0^{e^{-1}} \frac{dr}{r(-\log r)^{3\beta/2}} = 4\pi \int_1^\infty \frac{du}{u^{3\beta/2}} .$$

B. *Rollnik Potentials not in* $L^{3/2}$

In an appendix to this section, we will show $|\hat{V}_\beta(p)| < C|p|^{-1} (\log p)^{-\beta}$ for p large. Since $\hat{V}_\beta$ is bounded, we then have $\int p^{-1}|\hat{V}_\beta(p)|^2\, d^3p < \infty$ if and only if $\int^\infty d^3p\, [p^{-3} (\log p)^{2\beta}] < \infty$; thus

Example 3. Let V_β be as in example 2. If $\beta > 1/2$, $V_\beta \epsilon R$. Thus if $\frac{2}{3} \geq \beta > 1/2$, $V_\beta \epsilon R$ but $V_\beta \notin L^{3/2}$.

Remark.

Examples of the V_β type exhibit explicitly the well-known fact that the Fourier transform maps L^p into L^q for

$$1 < p < 2 \ (q^{-1} + p^{-1} = 1)$$

but is not quite onto. For example $\hat{V}_\beta \epsilon L^3$ if and only if $\beta > \frac{1}{3}$. Thus for β $(\frac{1}{3}, \frac{2}{3}]$, $\hat{V}_\beta \epsilon L^3$ but $V_\beta \notin L^{3/2}$. By changing the power in r, we can arrange examples of $V \notin L^p$ with $\hat{V} \epsilon L^q$ for β in an interval $[q^{-1}, p^{-1}]$. As $p \uparrow 2$, this interval shrinks to the empty set (as it should since $\hat{}$ is onto L^2 from L^2!)

C. *Rollnik Potentials which are nowhere locally* L^2

Consider a potential $W(x) = |x|^{-3/2}$ for $|x| < 1$ and 0 otherwise. Then $W \notin L^2$ but $W \epsilon L^1 \cap L^{3/2}$. Let q_n by a counting of all points in R^3 with rational coordinates. Define:

$$V(\vec{x}) = \sum_{n=1}^{\infty} 2^{-n}\, W(\vec{x} - \vec{q}_n) \cdot \tag{I.20}$$

Since $\Sigma 2^{-n}\|W(\cdot - q_n)\|_a < \infty$ for $a = 1, 3/2$, we see that the sum defining V converges a.e. to a function $V \epsilon L^1 \cap L^{3/2}$. But V is not L^2 over any sphere, for any sphere must contain some q_n and $V(r) \geq 2^{-n} W(r-q_n)$ is not L^2 around q_n. Thus:

Example 4: V as defined by (I.20) is in $L^1 \cap R$ but is nowhere locally L^2. Thus no continuous function is in D(V) and, in particular, $D(V) \cap D(H_0) = \{0\}$ since every function in $D(H_0)$ is continuous.[17]

Remark.

By taking $W(\vec{r}) = |r - r_0|^{-1/2}$ and summing over suitable $r_0 \epsilon R$, we find a central potential, V, with the properties of Example 4.

D. *Yukawian Potentials*

We recall that a Yukawian potential is a function V(r) given by:

$$V(r) = r^{-1} \int_{\mu_0}^{\infty} e^{-r\mu} \sigma(\mu)\, d\mu \qquad \text{(I.21)}$$

where σ is a tempered distribution with support in $[\mu_0, \infty)$. Let us consider the case where $\sigma(\mu)\, d\mu$ is a positive measure (in particular, σ could be a positive function). Then:

THEOREM I.27. *A Yukawian potential of form* (I.21) *with* $\sigma(\mu)$ *a positive measure has* $V \epsilon R$ *if and only if:*

$$\int_{\mu_0}^{\infty} d\mu\, \sigma(\mu) \int_{\mu_0}^{\infty} d\nu\, \sigma(\nu) \frac{\log \mu - \log \nu}{\mu^2 - \nu^2} < \infty \; .$$

Proof. Since everything is positive, we interchange order of integration freely. Thus, this theorem is a consequence of the computation in the appendix to this section where we show:

$$\int d^3x \int d^3y \frac{e^{-\mu|x|}\, e^{-\nu|y|}}{|x|\, |y|\, |x-y|} = (\text{constant}) \frac{\log \mu - \log \nu}{\mu^2 - \nu^2} . \; \blacksquare$$

---

17 See Kato [T17] p. 301.

COROLLARY I.28. *If* V *is a Yukawian potential with* $\sigma(\mu)\,d\mu$ *a measure, then a sufficient (but not necessary) condition for* $V \in R$ *is:*

$$\int_{\mu_0}^{\infty} d\mu\,|\sigma(\mu)| \int_{\mu_0}^{\infty} d\nu\,|\sigma(\nu)|\, \frac{\log\mu - \log\nu}{\mu^2 - \nu^2} < \infty .$$

Appendix to I.6: *Evaluation of Some Integrals*

(*a*) The Fourier transform of V_β ($\beta < 1$) (Examples 2, 3)

$$\hat{V}_\beta(p) = \int_0^{e^{-1}} \frac{r^2\,dr}{r^2(-\log r)^\beta} \int d\Omega\, e^{i\vec{p}\cdot\vec{r}}$$

$$= \frac{4\pi}{p} \int_0^{e^{-1}} \frac{dr}{r(-\log r)^\beta} \sin(pr)\,. \tag{I.22}$$

If we change variables to $x = pr$, we see

$$\hat{V}_\beta(p) = \frac{4\pi}{p} \int_0^{e^{-1}p} \frac{\sin x}{x} \frac{dx}{[\log p - \log x]^\beta}$$

which formally falls as $4\pi\, p^{-1}(\log p)^{-\beta} \int_0^\infty x^{-1} \sin x\, dx$ asymptotically. To justify Example 3, we need only to prove the weaker result $|\hat{V}_\beta(p)| < Cp^{-1}(\log p)^{-\beta}$ for p large. Dropping the $4\pi/p$, we write (I.22) as $\int_0^{p^{-1}} + \int_{p^{-1}}^{e^{-1}} = \mathrm{I} + \mathrm{II}$. Then $\mathrm{I} = \int_0^1 dx\,(\log p - \log x)^{-\beta} x^{-1}$

$\sin x < \int_0^1 du\,|\log p|^{-\beta} < (\log p)^{-\beta}$, and

$$\mathrm{II} = \int_{p^{-1}}^{e^{-1}} dx\, x^{-1}(-\log x)^\beta (-p^{-1}) \frac{d}{dx}(\cos px) = \mathrm{III} + \mathrm{IV}$$

where III = boundary term from an integration by parts =

$-p^{-1} \Big|_{e^{-1}}^{p^{-1}} x^{-1}(-\log x)^{-\beta}(\cos px)$ is bounded by $(pe^{-1})^{-1} + (\log p)^{-\beta}$

is bounded by $C(\log p)^{-\beta}$ for p large. Finally:

$$IV = p^{-1} \int_{p^{-1}}^{e^{-1}} dx\ (\cos px)\ \frac{d}{dx}\ [x^{-1}(-\log x)^{-\beta}]$$

$$|IV| \le p^{-1} \int_{p^{-1}}^{e^{-1}} dx\ \left| \frac{d}{dx}\ [x^{-1}(-\log x)^{-\beta}]\right| .$$

But $x(-\log x)^{\beta}$ has a positive derivative for $\log x < -\beta$ (and so for $x \epsilon [p^{-1}, e^{-1}]$). Thus IV is bounded by $(pe^{-1})^{-1} + (\log p)^{-\beta}$ which completes the proof.

(b) The R-norm of a Yukawa.

It is well known that the Fourier transform (up to a constant independent of μ) of $e^{-\mu r}/r$ is $(p^2 + \mu^2)^{-1}$. Thus by (I.8):

$$\int \frac{e^{-\mu|x|}}{|x|} \frac{e^{-\nu|x|}}{|y|} \frac{d^3x\ d^3y}{|x-y|^2} = C \int \frac{d^3p}{p} \frac{1}{p^2+\mu^2} \frac{1}{p^2+\nu^2}$$

$$= C' \int_0^\infty \frac{p\ dp}{(p^2+\mu^2)\ (p^2+\nu^2)}$$

$$= C'' \int_0^\infty \frac{d\kappa}{(\kappa+\mu^2)\ (\kappa+\nu^2)}$$

$$= C'' \lim_{R\to\infty} \int_0^R \frac{d\kappa}{\mu^2-\nu^2} \left[\frac{1}{\kappa+\nu^2} - \frac{1}{\kappa+\mu^2}\right]$$

$$= 2C'' \frac{\log\mu - \log\nu}{\mu^2-\nu^2} .$$

CHAPTER II

THE HAMILTONIAN

II.1. *Introduction*

From a fundamental point of view, the basic dynamical object in a quantum mechanical system is the propagator, U(t), which is a one parameter group of unitary operators.[1] On the other hand, the objects one obtains from classical mechanics are the Hamiltonian and its dynamics as determined by the Schrödinger equation. Stone's theorem (Theorem A.21)[2] provides the connection between the two approaches; on its authority we know that there is a one-to-one correspondence between unitary one-parameter groups and self-adjoint operators via:

$$U(t) = e^{-iHt}$$

The connection with Schrödinger's equation is provided by:

$$i\frac{d}{dt}(U(t)\Psi) = HU(t)\Psi$$

for $\Psi \epsilon D(H)$.

Since Stone's theorem deals with self-adjointness rather than mere Hermiticity, a mathematically non-trivial question immediately arises: to

---

1 Actually, the fundamental object is a group of automorphisms of the rays. However, an analysis of Wigner and Bargmann [5, 6, 123, T28] allows one to show that the automorphisms have a realization as a *continuous* one parameter group of unitary operators.

2 Assorted mathematical theorems and definitions are collected in an appendix at the end of the monograph.

show that in some sense the Hamiltonian is self-adjoint. One[3] of the simplest and most elegant results along these lines is:

THEOREM II.1 (Kato [65]). *Let* $V \epsilon L^2(R^3) + L^\infty(R^3)$. *Then* $D(V) \supset D(H_o)$ *and the operator* $H = H_o + V$ *with domain* $D(H) = D(H_o) \cap D(V) \equiv D(H_o)$ *is self-adjoint.* ∎

This theorem depends on the Kato-Rellich theorem (Theorem A.7) and the simple fact that $V \epsilon L^2$ is Kato tiny relative (Definition A.14) to H_o.[4] This last observation has a proof analogous tc our discussion of Theorem I.21:

LEMMA II.2. *If* $V \epsilon L^2 (R^3)$, *then* $V \underset{T.K.}{<<} H_o$.

Proof.[5] $V(E-H_o)^{-1}$ has a kernel $- V(x)\ (4\pi|x-y|)^{-1}\ e^{-\kappa|x-y|}$ where $E = -\kappa^2$. Since $V \epsilon L^2$, this kernel is Hilbert-Schmidt (Appendix, part (d)) with Schmidt norm $\|V\|_2\ (8\pi\kappa)^{-1}$. Thus, for any a, we can find E with $\|V(E-H_o)^{-1}\| < a$, so $V \underset{T.K.}{<<} H_o$. ∎

In this chapter, we discuss a self-adjoint Hamiltonian, $H_o + V$, when $V \epsilon R + L^\infty$. By Corollary I.2, this class includes the class covered by Kato.

Before we can even consider the problem of self-adjointness of H, we must look at a problem that does not arise in the Kato case. For $V \epsilon L^2 + L^\infty$, we had $D(V) \supset D(H_o)$ so that $H_o + V$ had a natural definition on the dense set $D(H_o)$. But for $V \epsilon R$ we can sometimes have $D(V) \cap D(H_o) = \{0\}$ (Section I.6 C) and thus we must first face up to

---

3 Other approaches to the problem are reviewed in Section 2 of [71].

4 The L^∞ part of V defines a bounded operator which is always Kato tiny.

5 For the usual proof, see [T17] pp. 301-302.

finding a definition for H. The solution is quite simple; while $D(V) \cap D(H_0)$ may be small, we have $Q(V) \cap Q(H_0) = Q(H_0)$.[6] Thus, we can define H as a quadratic form with $Q(H) = Q(H_0)$. Then, using the smallness of $V^{1/2}$ relative to $H_0^{1/2}$, we can prove the self-adjointness of the operator associated with the form $H_0 + V$. The result is an operator whose physics we discuss in later chapters.

If one views the domain of an operator as absolutely fundamental, then the following question naturally arises: the Hamiltonian H is an extension of the operator $H_0 + V$ defined on $D(H_0) \cap D(V)$; in what sense is it the natural one? The best answer to such a question is to declare it irrelevant. In a sense that we will discuss below, $Q(H)$ is the fundamental physical object, not $D(H)$, and so the question of naturalness only arises when H cannot de defined in terms of a quadratic form on $Q(H_0) \cap Q(V)$. Nevertheless, we can take such a question seriously and answer it by saying:

(1) The Feynman path integral in many cases can be shown to converge (Section II.6) to the propagator defined by our extension H.

(2) In case $D(H_0) \cap D(V)$ is dense, our choice of extension for $H_0 + \lambda V$ is the unique choice which is analytic at $\lambda = 0$. (Section II.8).

(3) If we approximate V by Kato potentials, then $H_0 + V_n \to H_0 + V$ in the generalized sense whenever $V_n \to V$ in Rollnik norm (Section II.4).

(4) If $V \in L^1 \cap R$, and $V_n \in L^1 \cap L^2$ converge in $L^1 - R$ norms to V, then with our choice of H, the corresponding S-matrices converge strongly (Section IV.3).

We stated above that $Q(H)$ is physically more fundamental than $D(H)$. The argument for this claim has been made forcefully by Nelson [T23]. He

---

6 For definition of Q and other technical objects, see the Appendix, in particular, Def. A.25.

simply remarks that physically the expectation value of an operator is fundamental and that Q(H) is the set of those vectors for which the energy has finite expectation value, while D(H) is the set of those vectors for which the energy squared has finite expectation value. Since the energy is more fundamental than its square, Q(H) is more fundamental than D(H).

Given that one wishes to study Hamiltonians defined by quadratic forms, we can also ask to what extent the Rollnik class is natural. For comparison, we note that if D(H) is viewed as important, the Kato potentials (i.e. $V \epsilon L^2 + L^\infty$) are natural in the following precise sense:

THEOREM II.3. *Let* $V(\vec{r})$ *be a function with* $\overline{\lim_{r\to\infty}} |V(\vec{r})| < \infty$. *If* $D(V) \supset D(H_0)$, *then* $V \epsilon L^2 + L^\infty$. *If moreover* $\overline{\lim_{r\to\infty}} |V(\vec{r})| = 0$, *then the* L^∞ *part may be chosen with arbitrarily small* $\| \|_\infty$, *i.e.* $V \epsilon L^2 + (L^\infty)_\varepsilon$. (*Recall* $V+(W)_\varepsilon = \{f | (\forall\varepsilon)\ \exists v \epsilon V,\ w \epsilon W \text{ so } f = v + w,\ \|w\|_W < \varepsilon\}$).

Proof. Let $\overline{\lim} |V(\vec{r})| = N$. Let $V_2(\vec{r}) = V(\vec{r})$ when $|V(\vec{r})| < 2N$ and 0 otherwise; $V_1 = V - V_2$. Then $V_2 \epsilon L^\infty$. To show $V_1 \epsilon L^2$, we first note that V_1 vanishes outside some large sphere. There exists $\psi \epsilon D(H_0)$ which is identically 1 on such a sphere; thus $\psi \epsilon D(V)$ implies

$$\int d^3r\ |V\psi|^2 \geq \int |V(\vec{r})|^2\ d^3r < \infty .$$

The second part of the theorem is proven similarly. ∎

Note. Without any assumption at ∞, one can show that any V with $D(V) \supset D(H_0)$ is "uniformly locally L^2," see e.g. Nelson [T23].

When combined with Lemma II.2, this theorem implies the rather surprising fact that a *local* potential bounded near infinity is Kato tiny if we merely know $D(V) \supset D(H_0)$. This fact depends crucially on the result that functions in $D(H_0)$ are bounded (since this is the real reason $V \epsilon L^2$

implies $V \underset{\text{T.K.}}{<<} H_o$; see Kato [T17] pp. 301-302), and thus on the fact that the underlying space has dimensions less than 4 (i.e. ≤ 3). An analogous result is false in higher dimensions:

Example 1. $V = cr^{-2}$ in dim bigger than 4. (i.e. ≥ 5). This is Kato small by a result of Strichartz [106, 107][7] but it is not Kato tiny by an argument analogous to Proposition II.4 below.

When we regard Q(H) as fundamental, the situation is not as clean, essentially because in this case 3 is too large a space dimension (i.e. in one dimension, $Q(H_o)$ contains only bounded functions and so $Q(V) \supset Q(H_o)$ if and only if V is uniformly locally L^1 if and only if $|V|^{1/2} \underset{\text{T.K.}}{<<} H_o^{1/2}$). In order to salvage something, we restrict ourselves to non-nasty potentials. To see that $Q(V) \supset Q(H_o)$ does not imply $V \in R + L^\infty$, we consider some examples:

Example 2. $V = 1/8r^{-2}$. By a result in Courant and Hilbert [T6] pp. 446-447, for any $\psi \in C_o^\infty$, $\|\frac{1}{2r}\psi\| \leq \|H_o^{1/2}\psi\|$. Since C_o^∞ is a core for $H_o^{1/2}$, this inequality carries over to all of $D(H_o^{1/2})$. Thus $D(H_o^{1/2}) \subset D(|V|^{1/2})$ and $|V|^{1/2} \underset{\text{T.K.}}{<<} H_o^{1/2}$. On the other hand, it is false in this case that $|V|^{1/2} \underset{\text{T.K.}}{<<} H_o^{1/2}$, for:

PROPOSITION II.4. If $\|\frac{1}{r}\psi\| \leq a\|H_o^{1/2}\psi\| + b\|\psi\|$ for all $\psi \in D(H_o^{1/2})$, then the same inequality holds with $b = 0$. Thus, there is a smallest a for which such an inequality holds.

---

7 This result says that $\|V\psi\| < a\|H_o\psi\| + b\|\psi\|$ for some fixed a and b and all $\psi \in D(H_o)$ whenever $V \in [L^{n/2}(R^n)]_W$, $n \geq 5$.

Proof. Let $\psi_n(x) = \psi(nx)$. Then $\|\psi_n\| = n^{-3/2}\|\psi\|$, $\|r^{-1}\psi_n\| = n^{-1/2}\|r^{-}$

$\|H_0^{1/2}\psi_n\| = n^{-1/2}\|H_0^{1/2}\psi\|$ so putting ψ_n into the inequality yields

$$\|r^{-1}\psi\| \leq a\|H_0^{1/2}\psi\| + bn^{-1}\|\psi\| .$$

Thus the inequality holds with $b = 0$. ∎

Consequently, we have a counterexample to a result analogous to Theorem II.3, even for monotone central potentials.

Example 3. $V_\alpha = (-\log r)^{-\alpha}r^{-2}$; $r < e^{-1}$). As we have seen, $V \epsilon R$ only when $\alpha > \frac{1}{2}$. On the other hand, for any $\alpha > 0$, $|V|^{1/2} \underset{\text{T.K.}}{<<} H_0^{1/2}$; for given α, a, pick R_0 so that $|V_\alpha(r)| < \frac{a}{4}r^{-2}$ for $r < R_0$. Then $|V_\alpha(r)| < \frac{a}{4}r^{-2} + C_{a,\alpha}$ for all r and thus

$$\|\, |V|^{1/2}\psi\| \leq a\|H_0^{1/2}\psi\| + C\|\psi\| .$$

However, (as might be expected from Chapter I) it is merely a question of logs:

THEOREM II.5. *Let* V *be a non-nasty potential. Then the following are equivalent:*

(1) $D(|V|^{1/2}) \supset D(H_0^{1/2})$.

(2) $V \epsilon (L^{3/2})_W + L^\infty$.

(3) $|V(r)| \leq Cr^{-2} + d$; all r.

(4) $\lambda|V|^{1/2} < H_0^{1/2}$ for some λ.

Proof. We consider (for simplicity) central monotone potentials and prove (1) → (2) → (3) → (4) → (1).

(1) → (2). If (2) does not hold, $(\forall n)\,(\forall T)\,(\exists t > T)$ with $r_t > nt^{-1/2}$. Pick some T_0 and choose $t_n > T_0$ so $r_{t_n} \equiv r_n > nt_n^{-1/2}$. Let $\psi \geq 0$ be a function in C_0^∞ which is identically 1 on the unit sphere and let

$\psi_n(r) = \psi(r/r_n)$. Then $\|\psi_n\| = Cr_n^{3/2}$, $\|H_0^{1/2}\psi_n\| = C'r_n^{1/2}$ and

$$\int |V(x)|\ |\psi_n(x)|^2 \geq t_n(r_n)^3 \geq n^2 r_n.$$

Consider the formal sum $\phi = \Sigma\, r_n^{-1/2} n^{-3/2} \psi_n$. Since

$$\Sigma\, r_n^{-1/2} n^{-3/2} \|\psi_n\| \leq r_{T_0}\ C\, \Sigma\, n^{-3/2} < \infty$$

and $\Sigma\, r_n^{-1/2} n^{-3/2} \|H_0^{1/2}\psi_n\| < C'\, \Sigma\, n^{-3/2} < \infty$, $\phi \epsilon D(H_0^{1/2})$. On the other hand,

$$\int |V(x)|\ |\phi(x)|^2 \geq \sum_n \int |V(x)|\ |\psi_n(x)|^2\, r_n^{-1} n^{-3} \geq \sum_n n^{-1} = \infty$$

so $\phi \notin D(|V|^{1/2})$. This contradicts (1) and thus (2) holds.

(2) → (3) As in Theorem I.20.

(3) → (4) Follows from Proposition II.4.

(4) → (1) Immediate. ∎

Thus, at least for monotone potentials, if $Q(V) \equiv D(|V|^{1/2}) \supset Q(H_0)$, we have at worst V barely not in $R + L^\infty$ (and, e.g. $[1 + |\log r|]^{-1}\, V(r) \epsilon R + L^\infty$).

Theorem II.5 leads one to feel that for some purposes a slightly larger class than $R + L^\infty$ is more natural. However, when V is not non-nasty, this class doesn't seem to have a simple direct characterization. Moreover, when we study scattering theory, we will see R arising quite naturally. We thus avoid introducing additional classes of potentials and study the classes $R + L^\infty$, $R + (L^\infty)_\varepsilon$, R and $R \cap L^1$ only.

After defining the Hamiltonian by a forms method in II.2, and alternately by Neumann (and summed Neumann) series in II.3, we investigate various general properties of these Hamiltonians. In this chapter, we treat mainly those features not directly related to spectra; in Chapter III, we turn to the bound state spectrum and in Chapter IV to the continuous spectrum and the related problem of scattering.

II.2. *Definition as a Quadratic Form*

In this section we will define the Hamiltonian $H_o + V$ for $V \epsilon R + L^\infty$ by a method closely related to the "Friedrichs extension." The idea of extending semi-bounded operators is due to von Neumann [92]; the existence of a natural extension was proved by Stone [T26] and Friedrichs [37]; various formal simplifications may be found in the work of Freudenthal [36], Calkin [11], Eberlein [28], Lions [T19] and Nelson [T22]. We use explicitly the idea of a scale of spaces. Faris has used this method to define $H_o + V$ for $V \epsilon L^{3/2}$. Simultaneous to the preparation of this work, Nelson has discussed this latter case in more detail [T23]. Kato, in his book [T17], has some discussion of the use of Freidrichs extensions in defining quantum mechanical Hamiltonians.

DEFINITION. Let $\mathcal{H}_{+1} = D(H_o^{1/2})$, endowed with the Hilbert space structure given by

$$h_{+1}(\Psi,\Phi) = \langle H_o^{1/2}\Psi, H_o^{1/2}\Phi\rangle + \langle\Psi,\Phi\rangle$$

$$= \langle\Psi, (H_o + 1)\Phi\rangle \,.$$

We will denote $\sqrt{h_{+1}(\Phi,\Phi)}$ by $\|\Phi\|_{+1}$. It is clear that $\|\Phi\| \le \|\Phi\|_{+1}$ for $\Phi \epsilon \mathcal{H}_{+1}$.

By the Riesz lemma, we can associate $\mathcal{H}_{+1}$ and the space $\mathcal{H}_{-1}$ of continuous conjugate linear functions on $\mathcal{H}_{+1}$. *We suppress this association* and regard $\mathcal{H}_{-1}$ as a separate space with norm $\| \ \|_{-1}$. Let $\Psi \epsilon \mathcal{H}$ (i.e. the original space $L^2(R^3)$). Then $\Phi \leadsto \langle\Phi,\Psi\rangle$ defines a conjugate linear function which is $\| \ \|_{+1}$ continuous since

$$|\langle\Phi,\Psi\rangle| \le \|\Phi\| \, \|\Psi\| \le \|\Phi\|_{+1} \, \|\Psi\|. \tag{II.1}$$

Thus we can regard $\mathcal{H} \subset \mathcal{H}_{-1}$ and with this association $\|\Psi\| \ge \|\Psi\|_{-1}$ (by (II.1)). In particular, $\mathcal{H}_{+1} \subset \mathcal{H}_{-1}$ and this inclusion is distinct from the natural map which we have suppressed. If $\Psi \epsilon \mathcal{H}$, we have:

$$\|\Psi\|_{-1} = \sup_{0 \neq \Phi \epsilon \mathcal{H}_{+1}} |\langle\Psi,\Phi\rangle| / \|\Phi\|_{+1}$$

$$= \sup_{0 \neq \Lambda \epsilon \mathcal{H}} |\langle\Psi, (H_0+1)^{-1/2}\Lambda\rangle| / \|\Lambda\|$$

$$= \|(H_0+1)^{-1/2}\Psi\|$$

$$= \sqrt{\langle\Psi, (H_0+1)^{-1}\Psi\rangle} .$$

This explains the notation $\mathcal{H}_{-1}$.

To summarize, we have introduced a pair of abstract Hilbert spaces along with $\mathcal{H}$, so that

$$\mathcal{H}_{+1} \subset \mathcal{H} \subset \mathcal{H}_{-1}$$

and so that

$$\|\Phi\|_{+1} \geq \|\Phi\| \ (\Phi \epsilon \mathcal{H}_{+1}); \quad \|\Phi\| \geq \|\Phi\|_{-1} \ (\Phi \epsilon \mathcal{H}).$$

Moreover, there is a "pairing" between $\mathcal{H}_{+1}$ and $\mathcal{H}_{-1}$ (given by evaluating the functional in $\mathcal{H}_{-1}$), which we represent by $\langle , \rangle_p$ so that for $\Psi \epsilon \mathcal{H}$, $\Phi \epsilon \mathcal{H}_{+1}$;

$$\langle\Phi,\Psi\rangle_p = \langle\Phi,\Psi\rangle .$$

We can thus drop the p in $\langle , \rangle_p$ without creating confusion. We then have:

$$|\langle\Phi,\Psi\rangle| \leq \|\Phi\| \ \|\Psi\| \quad \text{if } \Phi,\Psi \epsilon \mathcal{H}$$

$$|\langle\Phi,\Psi\rangle| \leq \|\Phi\|_{+1} \ \|\Psi\|_{-1} \quad \text{if } \Phi (\mathcal{H}_{+1}, \ \Psi \epsilon \mathcal{H}_{-1} .$$

Since $\overline{\langle\Phi,\Psi\rangle} = \langle\Psi,\Phi\rangle$ whenever both sides make sense, we can define $\langle\Psi,\Phi\rangle$ when $\Phi \epsilon \mathcal{H}_{+1}$, $\Psi \epsilon \mathcal{H}_{-1}$ by formally extending this relation.

We remark in passing that in the same way that $\mathcal{H}_{+1} \subset \mathcal{H}$ extended to $\mathcal{H}_{+1} \subset \mathcal{H} \subset \mathcal{H}_{-1}$, we can extend this to:[8]

$$\dots \mathcal{H}_{n+1} \subset \mathcal{H}_n \subset \dots \subset \mathcal{H}_{+1} \subset \mathcal{H} \subset \mathcal{H}_{-1} \subset \dots \subset \mathcal{H}_{-n} \subset \mathcal{H}_{-n-1} \subset \dots$$

where $\mathcal{H}_m$ has the norm:

$$\|\Psi\|_m = <\Psi, (H_0+1)^m \Psi>.$$

These spaces are nothing but the familiar Sobolev spaces (see e.g. [T29], pg. 55). We can picture them explicitly in p-space as those locally L^2 functions for which $(1+p^2)^{m/2} \psi(p) \epsilon L^2$. From this realization, we see that $m > n$ implies $\mathcal{H}_m$ is $\| \; \|_n$-dense in $\mathcal{H}_n$; in particular, $\mathcal{H}_{+1}$ is $\| \; \|$-dense in $\mathcal{H}$ and $\| \; \|_{-1}$ dense in $\mathcal{H}_{-1}$.

The crucial fact we will use about the scale $\mathcal{H}_{+1} \subset \mathcal{H} \subset \mathcal{H}_{-1}$ is the following:

LEMMA II.6. *Let* A_0 *be a continuous map from* $\mathcal{H}_{+1} \to \mathcal{H}_{-1}$, *such that* $A_0 + \lambda$ *is a bijection for some real* λ *and so that* $<\Psi, A_0\Phi> = <A_o\Psi,\Phi>$ *for all* $\Phi,\Psi \epsilon \mathcal{H}_{+1}$. *Then there is a unique self-adjoint operator A (on* $\mathcal{H}$*) so that* $D(A) \subset \mathcal{H}_{+1}$ *and* $<\Phi,A\Psi> = <\Phi,A_0\Psi>$ *for all* $\Psi \epsilon D(A)$ *and* $\Phi \epsilon \mathcal{H}_{+1}$.

Proof. We need only prove existence for the case $\lambda = 0$, for, in general, setting $B_0 = A_0+\lambda$ and $A = B - \lambda$, we recover the full theorem. We define D(A) by $D(A) = \{\Psi \epsilon \mathcal{H}_{+1} | A\Psi \epsilon \mathcal{H}\}$ and $A = A_0 \upharpoonright D(A)$. To show A is self-adjoint we consider the operator $A_0^{-1} \upharpoonright \mathcal{H}$. This operator is Hermitean for let $\Psi_1,\Psi_2 \epsilon \mathcal{H}$ and let $\Psi_i = A_0\Phi_i$, $\Phi_i \epsilon \mathcal{H}_{+1}$ (A_0 is onto). Then:

$$<\Psi_1, A_0^{-1}\Psi_2> = <A_o\Phi_1,\Phi_2> = <\Phi_1,A_0\Phi_2> = <A_o^{-1}\Psi_1,\Psi_2>.$$

---

8 One also defines $\mathcal{H}_\infty = \bigcap_{n=+\infty}^{-\infty} \mathcal{H}_n$; $\mathcal{H}_{-\infty} = \bigcup_{n=-\infty}^{\infty} \mathcal{H}_n$.

As a result $A_0^{-1} \restriction \mathcal{H}$ is an everywhere defined Hermitean operator and thus, by the Hellinger-Toeplitz theorem (Theorem A.2), it is bounded and self-adjoint. Therefore, A, as the inverse of $(A_0^{-1} \restriction \mathcal{H})$ is self-adjoint. Thus an A obeying the theorem exists. Let $\tilde{A}$ be another self-adjoint operator with $D(\tilde{A}) \subset \mathcal{H}_{+1}$ and $\langle\Phi,\tilde{A}\Psi\rangle = \langle\Phi,A_0\Psi\rangle$. Then A is an extension of $\tilde{A}$ for $\Psi \in D(A)$ implies $A_0\Psi = \tilde{A}\Psi \in \mathcal{H}$. Thus $\tilde{A} = A$. ∎

Note. For a brief history of results related to this theorem, see Cannon [16], pg. 57; Cannon calls this result the "KLMN theorem" for Kato, Lions, Lax, Milgram, Nelson.

This lemma allows us to prove the analogue of the Kato-Rellich theorem for forms:[9]

THEOREM II.7. *Let* V *define a* $\langle\ ,\ \rangle$ *symmetric form on* $\mathcal{H}_{+1}$ *with* $|V|^{1/2} \underset{\text{T.K.}}{<} H_0^{1/2}$, *for some* a < 1, *and some* b,

$$|\langle\psi,V\psi\rangle| \leq a\,\langle\psi,H_0\psi\rangle + b\,\langle\psi,\psi\rangle \text{ all } \psi \in \mathcal{H}_{+1}\,. \qquad \text{(II.2)}$$

Then, there exists a unique self-adjoint operator H *with* $D(H) \subset \mathcal{H}_{+1}$ *and*

$$\langle\phi,H\psi\rangle = \langle\phi,H_0\psi\rangle + \langle\phi,V\psi\rangle, \textit{ all } \psi \in D(H);\ \phi \in \mathcal{H}_{+1}. \qquad \text{(II.3)}$$

Moreover $Q(H) = \mathcal{H}_{+1}$ *and*

$$\langle\phi,H\phi\rangle = \langle\phi,(H_0+V)\phi\rangle \textit{ all } \phi \in \mathcal{H}_{+1}.^{10} \qquad \text{(II.4)}$$

Proof. Define $V\colon \mathcal{H}_{+1} \to \mathcal{H}_{-1}$ by $\langle V\phi,\psi\rangle = V(\phi,\psi)$. To show V is continuous, we must work on extending (II.2). (II.2) implies

---

9 A similar result can be found in many places, e.g. Kato [T17] pp. 338-343 or Nelson [132], App.

10 This formula is intended in the sense of expectation values; $H\phi$ may not make sense as a bona fide vector in $\mathcal{H}$.

$|\langle\psi,V\psi\rangle| \leq C\|\psi\|_{+1}^2$ for some C. Let $h(\phi,\psi) = \langle\phi,(CH_o + C - V)\psi\rangle$. Then h is an Hermitean bilinear form with $0 \leq h(\phi,\phi) \leq 2C\|\phi\|_{+1}^2$ $(\phi \neq 0)$. Thus, h is an inner product and so a Cauchy-Schwartz inequality holds: $|h(\phi,\psi)|^2 \leq h(\phi,\phi)\, h(\psi,\psi) \leq 4C^2\, \|\phi\|_{+1}^2$. Thus:

$$|\langle\phi,V\psi\rangle| \leq |C\langle\phi,(H_o + 1)\psi\rangle| + |h(\phi,\psi)|$$
$$< 3C\, \|\phi\|_{+1}\, \|\psi\|_{+1}.$$

This implies V is continuous.

We next show that $H_o + V + b + 1$ is bijective as a map of $\mathcal{H}_{+1} \to \mathcal{H}_{-1}$. For (by (II.2)):

$$\langle\psi, (H_o + V + b + 1)\psi\rangle \geq (1-a)\, \langle\psi,(H_o+1)\psi\rangle = (1-a)\, \|\psi\|_{+1}^2. \quad \text{(II.5)}$$

Thus $(H_o + V + b + 1)\psi = 0$ implies $\|\psi\|_{+1} = 0$ implies $\psi = 0$ so $H_o + V + b + 1$ is injective. Also, by (II.5), Ran $(H_o + V + b + 1)$ is closed. If it is not all of $\mathcal{H}_{-1}$ we can find $\phi\epsilon[\text{Ran}\,(H_o + V + b + 1)]^{\perp} \subset \mathcal{H}_{+1}$. But then $\langle\phi, (H_o + V + b + 1)\phi\rangle = 0$ implies $\phi = 0$ (by (II.5)). Thus $H_o + V + b + 1$ is bijective so Lemma II.6 tells us that H obeying (II.3) can be defined.

To prove $Q(H) = \mathcal{H}_{+1}$, we use Theorem A.20, which tells us that $\psi\epsilon Q(H)$ if and only if $\psi_n \to \psi$ in $\|\ \|$ and $\langle(\psi_n - \psi_m), H(\psi_n - \psi_m)\rangle \to 0$ for some sequence $\psi_n \epsilon D(H)$. By (II.2) we have:

$$\langle\psi,H_o\psi\rangle \leq (1-a)^{-1}\, [\langle\psi,H\psi\rangle + b\, \langle\psi,\psi\rangle] \quad \text{(II.6)}$$

$$\langle\psi,H\psi\rangle \leq (1+a)^{-1}\, \langle\psi,H_o\psi\rangle + b\, \langle\psi,\psi\rangle. \quad \text{(II.7)}$$

---

11 Since $(\mathcal{H}_{+1})^* = \mathcal{H}_{-1}$ and $\mathcal{H}_{+1}$ is reflexive, $(\mathcal{H}_{-1})^* = \mathcal{H}_{+1}$.

Thus, if $\psi_n \to \psi$ we have $\langle(\psi_n - \psi_m), H(\psi_n - \psi_m)\rangle \to 0$ if and only if $\langle(\psi_n - \psi_m), H_0(\psi_n - \psi_m)\rangle \to 0$. As a result $Q(H) = Q(H_0) = \mathcal{H}_{+1}$. ∎

Remarks.

1. While this theorem is analogous to the Kato-Rellich theorem, its proof is completely different. In particular, in proving this result from first principles, no simplification results by introducing the idea of a graph.

2. (II.6) and (II.7) imply that for E sufficiently large, the norms $\| \ \|_{+1}$ and $\psi \to (\langle\psi,(H+E)\psi\rangle)^{1/2}$ are equivalent. This remark will play a role in many of the arguments to follow.

3. The form of H_0 was not essential. H_0 bounded below was all that was used.

4. We remark that $D(H_0) \cap D(V) \subset D(H)$. So H is an extension of $H_0 + V$.

COROLLARY II.8. *Let* $W \in R + L^\infty$. *Then there is a unique self-adjoint operator* H *with* $Q(H) = Q(H_0)$ *and* $\langle\psi,H\psi\rangle = \langle\psi,H_0\psi\rangle + \langle\psi,W\psi\rangle$ *all* $\psi \in \mathcal{H}_{+1}$. *In case* $W = V + K$ *with* $K \in L^\infty$, $V \in L^1 \cap R$, *the domain of* H *can be described explicitly in any of the following ways:*

(a) *Let* $\psi \in D(H_0^{1/2})$. *View* ψ *as a distribution. If* $f \in C_0^\infty$, $V^{1/2} f \in L^2$ *and* $\|V^{1/2} f\|_2 \leq \|f\|_\infty \|V\|_1$. *If* $\psi \in D(H_0^{1/2})$, $|V|^{1/2}\psi \in L^2$ *and so* $f \leadsto \int f(V\psi) = \int (V^{1/2} f)(|V|^{1/2}\psi)$ *defines a distribution (for* $|\langle \overline{V^{1/2} f}, |V|^{1/2}\psi\rangle| \leq \||V|^{1/2}\psi\|_2 \|V\|_1 \|f\|_\infty$) *which we represent as* $V\psi$. *Let* $H_0\psi = -\Delta\psi$ *defined as distributional derivatives. Then* $\psi \in D(H)$ *if and only if* $H_0\psi + V\psi \in L^2$ *(as a distribution).*

(b) *Let* $\tilde{V}$ *by the Fourier transform of* V *(normalized so* $\widetilde{fg} = \tilde{f} * \tilde{g}$). *Let* $\psi \in D(H_0^{1/2})$ *and let* $\tilde{\psi}$ *be its Fourier transform. Then* $\psi \in D(H)$ *if and only if*

$$k^2\tilde{\psi}(k) + \int d^3p\, \tilde{V}(k-p)\, \tilde{\psi}(p) \in L^2(k).$$

(c) *In case* $D(H_0) \cap D(V)$ *is a core for* $H_0^{1/2}$, H *is just the Friedrich's extension of* $\tilde{H} = H_0 + V$ *defined on* $D(H_0) \cap D(V)$. *Then:*

$$D(H) = D([\tilde{H}]^*) \cap D(H_0^{1/2}).$$

Proof. (a) and (b) follow from the description of D(H) in Lemma II.6. (c) follows from a standard description of the Friedrich's extension; see e.g. [T29], pp. 317-318. ∎

Remark.

We will henceforth mean the operator H when we write $H_0 + V$.[12]

We will often use $|\langle\psi,\phi\rangle| \leq \|\psi\|_{+1}\ \|\phi\|_{-1}$ and so inequalities of the form $\|A\psi\|_{+1} \leq C\|\psi\|_{+1}$ will be very useful. We call the smallest such C, $\|A\|_{+1,+1}$. In fact:

THEOREM II.9. (a) $\|e^{iH_0 t}\|_{+1,+1} = 1$.

(b) $\|e^{iHt}\|_{+1,+1} \leq C$ where C is independent of t (but not of H).

Proof. (a) $\|e^{iH_0 t}\phi\|_{+1} = \langle e^{iH_0 t}\phi, (H_0+1)e^{iH_0 t}\phi\rangle = \|\phi\|_{+1}$.

(b) We have, following Theorem II.7, Remark 2:

$$\|\phi\|_{+1} \leq C_1 \langle\phi, (E+H)\phi\rangle \leq C_2\|\phi\|_{+1}\,.$$

Thus

$$\begin{aligned}\|e^{iHt}\phi\|_{+1} &\leq C_1 \langle e^{iHt}\phi, (H+E)e^{+iHt}\phi\rangle \\ &\leq C_1 \langle\phi, (H+E)\phi\rangle \leq C_2\|\phi\|_{+1}. \blacksquare\end{aligned}$$

Another technique we will use is to replace strong convergence on D(H) by weak convergence on Q(H); for example:

---

12 This is counter to the usual convention of defining A + B on $D(A) \cap D(B)$.

THEOREM II.10. (a) *If* $\psi, \phi \in \mathcal{H}_{+1}$, $\lim_{t\to 0} \langle\psi, t^{-1}(e^{iH_0 t}-1)\phi\rangle = i\langle\psi, H_0\phi\rangle$,

(b) *The same as* (a) *with* H_0 *replaced by* H.

Proof. Passing to a spectral representation, we have $\psi \equiv \psi(x)$, $\phi \equiv \phi(x)$ and $e^{iAt} \equiv$ multiplication by e^{ixt}. Then $\overline{\psi(x)}\,(e^{ixt}-1)\,t^{-1}\phi(x)$ converges pointwise to $i\,\overline{\psi(x)}\,x\,\phi(x)$. But $t^{-1}|e^{ixt}-1| < |x|$ so the functions $\overline{\psi(x)}\,(e^{ixt}-1)\,t^{-1}\phi(x)$ are dominated by an L^1-function (when $\psi, \phi \in Q(A)$). Thus the theorem follows from the dominated convergence theorem. ∎

II.3. *Tiktopoulos' Formula*

Tiktopoulos[13] has remarked that the formal expansion (for $E < 0$):

$$(H-E)^{-1} = (H_0-E)^{-1/2}\,[1 + (H_0-E)^{-1/2}\,V(H_0-E)^{-1/2}]^{-1}\,(H_0-E)^{-1/2} \tag{II.8}$$

can be used to define the Hamiltonian when $V \in R$,[14] for we have already seen that:

$$A_E = (H_0-E)^{-1/2}\,V(H_0-E)^{-1/2}$$

is a legitimate operator and $\mathrm{Tr}(A_E^{\dagger}A_E) \to 0$ as $E \to -\infty$ (Section I.4). Thus we can make sense out of the right hand of (II.8) by taking E so negative that $\|A_E\| \leq [\mathrm{Tr}(A_E^{\dagger}A_E)]^{1/2} < 1$. While we could have used (II.8) as a definition, we have already defined H by a quadratic form method, and thus we must show the two definitions are the same (or abandon (II.8)). We can summarize our discussion of the existence of the existence of the r.h.s. of (II.8) by:

---

13 Private communication.

14 By taking E very negative, we can use (II.8) even when $V \in R + L^{\infty}$; we consider R only, for simplicity's sake.

LEMMA II.10. *For* E *sufficiently negative,* $B_E = (H_0-E)^{-1/2}(1+A_E)^{-1}(H_0-E)^{-1/2}$ *exists and is given by* $B_E = \sum_{n=1}^{\infty}(-1)^{n+1}U_n$ *with*

$$U_n = (H_0-E)^{-1}V\ \dots\ V(H_0-E)^{-1}\quad (n-1\ V\text{'s}).\ \blacksquare$$

The expansion for B_E is $\|\ \|$-convergent. We must treat the expression for U_n with some care. Either we regard it as a formal expression for $(H_0-E)^{-1/2}(A_E)^n(H_0-E)^{-1/2}$, or we regard $(H_0-E)^{-1}: \mathcal{H} \to \mathcal{H}_{+1}$ and then $V: \mathcal{H}_{+1} \to \mathcal{H}_{-1}$, $(H_0-E)^{-1}: \mathcal{H}_{-1} \to \mathcal{H}_{+1}$, etc. We also remark that $\operatorname{Ran} B_E \subset \operatorname{Ran}(H_0-E)^{-1/2} = \mathcal{H}_{+1}$.

LEMMA II.11. *For* E *sufficiently negative,* $B_E = (H-E)^{-1}$.[15]

Proof. Treating everything as maps from $\mathcal{H}_{+1}$ to $\mathcal{H}_{-1}$ or vice-versa, we see that:

$$(H_0-E)^{-1}(H-E)\sum_{n=1}^{N}(-1)^{n+1}U_n = (H_0-E)^{-1} + (-1)^{N+1}U_{N+1}.$$

Apply both sides to a vector $\Psi \in \mathcal{H}$, we see (using $\|U_{N+1}\| \to 0$) that:

$$(H_0-E)^{-1}(H-E)B_E = (H_0-E)^{-1} = (H_0-E)^{-1}(H-E)(H-E)^{-1}$$

where we take E so negative that $(H-E)^{-1}$ exists. If we again view the maps as maps of the scale spaces (and use $\operatorname{Ker}(H-E) = \operatorname{Ker}(H_0-E)^{-1} = 0$ we see that $B_E\Psi = (H-E)^{-1}\Psi$ for any $\Psi \in \mathcal{H}$, $\blacksquare$

---

[15] H defined by a quadratic forms method.

We can now use analytic continuation to extend (II.8). Let us first define:

DEFINITION. The "canonically cut plane" means the complex plane with the non-negative reals removed.

THEOREM II.12. *Let* $V \in R$ *and let* $H = H_0 + V$ *be defined by the forms method. Then:*

(a) (II.8) *holds for all* E *in the canonically cut plane with the exception of a discrete set.*

(b) *For* E *sufficiently negative,*

$$(E-H)^{-1} = (E-H_0)^{-1} + (E-H_0)^{-1} V(E-H_0)^{-1} + \ldots\ldots \qquad \text{(II.9)}$$

Proof. A_E is compact for all E in the canonically cut plane and is analytic in E. Since $(1+A_E)^{-1}$ exists for E very negative $(1+A_E)^{-1}$ exists at all but a discrete set of points in the canonically cut plane and is analytic (Theorem A.27). Thus the right hand side of (II.8) is analytic off a discrete set. This implies the left hand side is analytic in the same region and so $(H-E)^{-1}$ exists in the same region. ∎

Remarks.

1. We will call (II.8), Tiktopoulos' formula.
2. (a) implies that the negative energy spectrum is pure point. We will later see it is actually finite when $V \in R$. This proof of the pure point nature is very close to the standard Weyl theorem proof in the Kato case [71].
3. We will see later (Section III.2) that (II.8) only fails when $E \in \operatorname{spec}(H)$.
4. (b) can easily be extended to the $R + L^\infty$ case and (a) to $V \in R + (L^\infty)_\varepsilon$.

II.4. *Approximation Theorems*

As an immediate consequence of Tiktopoulos' formula, (II.8), we have:

THEOREM II.13. *Let* V_n, $V \in R$ *and let* $H_n = H_0 + V_n$; $H = H_0 + V$ *as defined by the forms method. If* $V_n \to V$ *in Rollnik norm, then:*

$$(H_n - E)^{-1} \to (H-E)^{-1} \quad (\text{in } \|\ \|)$$

for any $E \notin \text{spec}(H)$.[16]

Proof. Let $A_E^{(n)} = (H_0 - E)^{-\frac{1}{2}} V_n (H_0 - E)^{-\frac{1}{2}}$ and

$$A_E = (H_0 - E)^{-\frac{1}{2}} V (H_0 - E)^{-\frac{1}{2}}.$$

Then

$$\text{Tr}\,[(A_E - A_E^{(n)})^\dagger (A_E - A_E^{(n)})] \leq \|V - V_n\|_R^2 \to 0 \ .$$

Thus, $E \notin \text{spec}(H)$ implies $(1 + A_E)^{-1}$ exists[16] implies

$$(1 + A_E^{(n)})^{-1} \to (1 + A_E)^{-1}.$$

Therefore, by Tiktopolous' formula, the result follows. ∎

We thus have convergence in the generalized sense and thereby according to the Trotter-Kato theorem (Theorem A.22):

COROLLARY II.14. *Let* $V_n \to V$ *in Rollnik norm. Then* $e^{-iH_n t} \to e^{iHt}$ *strongly.* ∎

We also have by Theorems (A.16) and (A.17), along with the discreteness of the negative energy spectrum proved in Section II.3, that:

COROLLARY II.15. *Let* $V_n \to V$ *in Rollnik norm. Let* $P_n(E)$, $P(E)$ *be the spectral projections for* H_n *and* H. *Then:*

(a) $P_n(\lambda,\infty) \to P(\lambda,\infty)$ *strongly when* $\lambda > 0$ *is not a positive energy eigenvalue (see Section III.4).*

---

16 We assume a result we won't prove until Chapter III, namely that $(1 + A_E)^{-1}$ fails to exist only when $E \in \text{spec}(H)$. Alternately, we could prove II.13 without this result by employing general properties of norm "generalized convergence."

(b) $P_n(a,b) \to P(a,b)$ *in* $\| \ \|$ *when* $a < b < 0$ *are not negative energy eigenvalues.* ∎

Remarks.

1. From the known discreteness of the negative energy spectrum for $V \epsilon L^2$ and Theorem II.13, one could provide an alternate proof of discreteness for $V \epsilon R$.

2. When $V \epsilon L^2$, we have:

$$(E-H)^{-1} = [1-(E-H_o)^{-1}V]^{-1} (E-H_o)^{-1} .$$

For E sufficiently negative. Thus, Theorem II.13 and its corollaries have analogues if $V_n \to V$ in L^2-norm (all V_n, $V \epsilon L^2$).

3. The convergence of the propagators is further evidence for the naturalness of our definition of H.

4. The C^∞ functions of compact support are dense in R and so we can approximate the physics with extremely nice potentials.

II.5. *Domains of Essential Self-Adjointness*

For any $V \epsilon R + L^\infty$, we have succeeded in defining a Hamiltonian H with $H \upharpoonright D(H_o) \cap D(V) \equiv H_o + V$ (in the ordinary sense of operator sums). In the general case, there is no hope that $D(H_o) \cap D(V)$ is a core (domain of essential self-adjointness, see Theorem A.5) since this subspace may not even be dense. However, in nice cases, one would hope to be able to show $D(H_o) \cap D(V)$ is a core and it is to this question that we next turn.

There is, of course, an enormous mathematical literature on domains of essential self-adjointness. There has been considerable work on the particular case of interest to quantum mechanics, especially on allowing misbehavior at ∞ (e.g. the harmonic and anharmonic oscillators, Stark effect, etc.). Besides the elementary results for $V \epsilon L^2 + L^\infty$ (in which case D is a core for H if and only if it is a core for H_o), there are three general techniques which appear to have been used:

(1) A technique going back to Carleman [17] which uses the regularity of weak solutions of elliptic equations; see also A. Jaffe [59, 60] and Wightman [121] pp. 268-269.

(2) General properties of differential operators are used; the best known result of this sort is due to Ikebe and Kato [57]; see also Kato [71] for additional references.

(3) Methods based on the Weyl limit point, limit circle method [119, 120] (see also [T5]). These methods are only applicable to one dimensional problems (see Wintner [125], Dunford and Schwartz [T7] pp. 1392-1435, Nelson [T23]) and central potentials (see Case [18] and Nelson [T23])

Using methods (2) and (3), "singular" potentials have been studied by various authors (method (2) by Jörgens [133]; method (3) by Case [18]; (see also Limić [131]) but these methods only apply to potentials more singular than r^{-2}. As a matter of fact we are very pessimistic about any general essential self-adjointness result holding for

$$r^{-\alpha},\ 2 > \alpha > 3/2,$$

for it is not hard to show using the Weyl limit point/limit circle case that on $C_0^\infty(R^3-\{0\})$, these operators have deficiency indices (1,1) [all in the subspace with $\ell = 0$]. Taking the form sum $H_0 + V$ in these cases is equivalent to taking the Friedrich's extension and these operators are presumably not essentially self-adjoint on $D(H_0) \cap D(V)$.

Note. Because of an incorrect theorem in the original manuscript, there are no Theorems II.16, II.17, or II.18. I should like to thank Prof. J. Walter for showing me the error.

II.6. *Feynman Path Integrals*

In 1948, Feynman [33] presented a physically appealing reworking of quantum mechanics in terms of a non-rigorous "path integral" (see also Feynman and Hibbs [T8]). The propagator $P(x,x';t)$ is defined so that:

$$\psi(x,t_0+t) = \int P(x,x';t)\,\psi(x',t_0)dx'. \tag{II.15}$$

$P(x,x';t)$ is given by an "integral" over paths $x(s)$ with $x(0) = x'$, $x(t) = x$ with integrand $\exp(i\,S[x(s)]/\hbar)$ where S is the classical action for the path. One of the beautiful consequences of this formulation is that in the classical limit where $S/\hbar$ varies very rapidly, the main contribution is from the region where the phase $S/\hbar$ is stationary, i.e. the path(s) of stationary action mainly contributes in the classical limit.

The earliest attempts at rigorizing the notion are due to Kac [64] who applied measure theory in "path space" to obtain the analogue of Feynman's formula for imaginary t (see also Kac [T16]). Other mathematical treatments and contributions are due to Babbitt [3], Cameron [15], Daletzski [25], Faris [29, 30], Feldman [32], Gel'fand and Yaglom [40], Ito [58], McShane [87] and Nelson [91]. One of the several points discussed by Nelson in [91] is the fact that the Trotter product formula (Theorem A.23) yields the Feynman path integral formula in the case where $V \in L^2 + L^\infty$. In fact, only a simple special case of Trotter's formula is needed.[17]

The Trotter formula requires $D(H_0) \cap D(V)$ to be a core for H, something unlikely unless V is in a Kato class; however Faris has proven:

LEMMA II.19. (Faris [29]). *Let* V *be self-adjoint and suppose:*

(a) $D(V) \cap \mathcal{H}_{+1}$ *is dense in* $\mathcal{H}_{+1}$.

(b) $V\colon D(V) \cap \mathcal{H}_{+1} \to \mathcal{H}_{-1}$ *extends to the continuous map* $V\colon \mathcal{H}_{+1} \to \mathcal{H}_{-1}$ *with norm strictly less than* 1. *Let* $t > 0$. *Then* $e^{iHt} = \lim_{\varepsilon\downarrow 0} \lim_{n\to\infty} [\exp(iH_0t/n(1-i\varepsilon)) \exp(iVt/n)]^n$ *where the limits are in the strong operator topology and* $H = H_0 + V$ *defined by a forms method.*

Proof. See Faris [29]. ∎

[17] The full Trotter theorem only requires $A + B$ to be essentially self-adjoint on $D(A) \cap D(B)$. The simpler case (for which Nelson provides an easy proof) requires $A + B$ to be self-adjoint on $D(A) \cap D(B)$.

Remarks.

1. Faris states his result in greater generality; we have abstracted the form needed.

2. This is not a direct extension of Trotter's theorem since the $\lim_{\varepsilon \downarrow 0}$ appears.

3. (b) is a statement that $V^{1/2} \underset{\text{T.K.}}{<} H_o^{1/2}$.[18]

4. Although Faris uses condition (a) in his proof, it has the "smell" of a disposable condition.

THEOREM II.20. *Let* $V \epsilon R + L^\infty$. *If* $D(V) \cap \mathcal{H}_{+1}$ *is a core for* $H_o^{1/2}$, *in particular, if* V *is locally* L^2 *in* R^3 *-finite set, then*

$$(e^{-iHt})(x) = \lim_{\varepsilon \downarrow 0} \lim_{n \to \infty} \left[\frac{4\pi it}{n(1+i\varepsilon)}\right]^{-3n/2} \int \cdots \int e^{iS_\varepsilon(x_o,\ldots,x_n;t)} \psi(x_n) dx_1 .. dx_n \qquad \text{(II.18)}$$

where

$$S(x_o,\ldots,x_n;t) = \sum_1^n \frac{(1+i\varepsilon)}{4} \frac{|x_j - x_{j-1}|^2}{(t/n)^2} - V(x_j) \frac{t}{n}. \qquad \text{(II.19)}$$ ∎

Remarks.

1. This is only the mildest of improvements of Faris' application which is stated for $V \epsilon L^{3/2} + L^\infty$.

2. If in fact condition (a) in Lemma II.19 is unneeded, then Theorem II.20 holds for any $V \epsilon R + L^\infty$. Until such a strengthened Lemma II.19 is proven or until $D(V) \cap \mathcal{H}_{+1}$ dense in $\mathcal{H}_{+1}$ is shown to be a weak condition we must regard the situation with respect to Feynman path integrals as unsatisfactory.

---

18 We may have to renorm $\mathcal{H}_{+1}$ with $\|\Psi\|'_{+1} = \|(H_o + E)^{1/2}\Psi\|$.

Note added in proof:

By a simple argument employing the spectral theorem for V, Faris has shown (unpublished) that Lemma II.19 holds with (a) and (b) replaced by $Q(V) \supset Q(H_0)$ and $V^{1/2} \underset{\text{T.K.}}{<} H_0^{1/2}$. Thus Theorem II.20 holds for any function $V \in R + L^\infty$, so the situation is satisfactory with respect to Rollnik potentials and Feynman integrals. It is a pleasure to thank Professor Faris for a valuable discussion.

II.7. *The Time-Dependent Schrödinger Equation*

In this section, we consider the definition of the propagator $U(t,s)$ when the Hamiltonian is time dependent.[19] We are then faced with attempting to solve a Schrödinger equation of the form

$$\frac{d\psi(t)}{dt} = -iH(t)\psi(t).$$

We seek a solution in the form of a "unitary propagator" where:

DEFINITION. We say a function $U(t,s)$ on $\mathbf{R} \times \mathbf{R}$ (or $[a,b] \times [a,b]$) is a unitary propagator if and only if:

(a) U is unitary.

(b) $U(t,s)\, U(s,v) = U(t,v)$.

(c) $U(t,t) = 1$.

(d) $U(t,s)^\dagger = U(s,t)$ (redundant!).

(e) $t,s \to U(t,s)$ is jointly continuous in the strong topology.

[19] One could argue that time dependent Hamiltonians only arise in the approximation of ignoring "recoil" effects on an external system; e.g. the motion of a particle in an external field is really generated by a time independent (field theoretic) Hamiltonian. However, since this approximation is believed to be extremely good, existence of solutions of the approximate equations seems to be an important question.

Time dependent equations (in the more general context of arbitrary Banach spaces with $-iH(t)$ the generator of a contraction semi-group) were first considered by Kato [66], who obtained a result requiring $D(H(t))$ to be time independent. Since then, an extensive literature has developed on the subject (see Yoshida [126, 127, 128], Lions [T19], Ladyzhenskaya–Visik [84], Kisynski [76]). In the Kato potential case[20] ($V \in L^2 + L^\infty$), one is able to use almost any of the theorems, e.g.:

THEOREM II.21 (Yoshida). *Let* $H(t)$ $(0 \leq t \leq 1)$ *be a time-dependent self-adjoint operator with* $H(t) \geq -E_o + 1$ *for some* E_o. *Suppose* $(H(t)+E_o)^{-1}$ *is strongly differentiable and* $\|(H(t)+E_o)\frac{d}{dt}[(H(t)+E_o)^{-1}]\|$ *is bounded. Then, for any* $x_o \in D(H(0))$, *there is a unique strongly continuous function* $x(t) \in D(H(t))$, $x(0) = x_o$ *and*

$$\frac{d}{dt}x(t) = -iH(t)x(t) \qquad \textit{(derivative in } \|\ \|\textit{-sense)}.$$

Moreover, $\|x(t)\| = \|x_o\|$. *Thus* $U(t,s)$ *defined by* $U(t,s)\,x(s) = x(t)$ *extends to all of* $\mathcal{H}$ *and defines a unitary propagator (on* $[0,1] \times [0,1]$).

Proof. See Yoshida [T29], pp. 425-429 (first edition!). ∎

Remarks.

1. It has been noted by Kato (see Yoshida [T29], p. 429) that $\|[H(t)+E]\frac{d}{dt}[(H(t)+E)^{-1}]\|$ bounded implies $D(H(t))$ is independent of time. Thus, this theorem is not suitable for Rollnik potentials.

2. If one makes use of the hermiticy of the $H(t)$'s, Yoshida's proof (which holds in any B-space) can be simplified.

---

20 Surprisingly, the literature does not appear to contain any explicit theorems expressed in terms of the potential $V(t)$. We thus first summarize the status of the Kato case.

One thereby concludes:

THEOREM II.22. *Let* $V(t) = V_1(t) + V_2(t)$ *be given* $(-\infty < t < \infty)$ *with* $V_1(t) \epsilon L^2(x); V_2(t) \epsilon L^\infty(x)$. *Suppose:*

(a) $V_1(t)$ *is differentiable in* t *as an* L^2*-valued function.*

(b) $V_2(t)$ *is differentiable in* t *as an* L^∞*-valued funtion.*

Let $H(t) = H_0 + V(t)$. *Then, for any* $\psi_0 \epsilon D(H_0)$, *there is a solution of* $\frac{d}{dt}\psi(t) = -iH(t)\psi(t);\ \psi(0) = \psi_0$. *The map* $\psi(s) \to \psi(t)$ *extends to a unitary propagator.*

Proof. Find $E > 0$, so that $H_0 + V(t) + E > 1$ all $t \epsilon [0,1]$. (Since $V_1(t)$ $[V_2(t)]$ is L^2 $[L^\infty]$ bounded on [0,1], this is possible.) Then:

$$(H(t) + E)^{-1} = (H_0 + E)^{-1} \ [1 + V(t) (H_0 + E)^{-1}]^{-1}.$$

By simple arguments, $V(t)(H_0 + E)^{-1}$ is bounded and has a derivative $\dot{V}(t)(H_0 + E)^{-1}$ in the bounded operator sense. Thus:

$$\begin{aligned}\frac{d}{dt}[(H(t) + E)^{-1}] &= (H_0 + E)^{-1} [1+V(t)(H_0+E)^{-1}]^{-1} \dot{V}(t)(H_0+E)^{-1} \\ &\qquad [1+V(t)(H_0+E)^{-1}]^{-1} \\ &= [H(t)+E]^{-1} [\dot{V}(t)(H_0+E)^{-1}] [1+V(t)(H_0+E)^{-1}]^{-1}\end{aligned}$$

so that $\|(H(t) + E)\frac{d}{dt}[(H(t) + E)^{-1}]\|$ is bounded. By Theorem II.21, solutions exist in [0,1] and similarly in [n,n+1] $(n=0,\pm 1,\ldots)$. We can piece together to obtain solutions on $(-\infty,\infty)$. ∎

Remarks.

1. $V_i(t)$ piecewise continuous with bounded derivatives in the intervals vals of continuity is sufficient for the theorem to hold (if we don't demand differentiability at the bad points).

2. Actually, smoothness of the L^∞ part is not needed; uniform boundedness is enough. For we can first solve $\frac{d\tilde{\psi}}{dt} = -i(H_o+V_1(t))\tilde{\psi}$ with $\tilde{U}(t,0)$. We then introduce an "interaction picture" $\psi_{int} = \tilde{U}(t,0)\psi$. Then $\frac{d\psi}{dt} = iH(t)\psi$ becomes $\frac{d\psi_{int}}{dt} = -i\ \tilde{V}_2(t)\psi_{int}$ where

$$\tilde{V}_2(t) = U(t,0)\ V_2(t)\ U(t,0)^\dagger$$

is uniformly bounded. This last equation can be solved by a convergent Dyson solution $U(t,0) = T(e^{i\int_0^t \tilde{V}_2(s)ds})$.

For the Rollnik case, we need a stronger result than Yoshida's; explicitly, we use the following two theorems:

THEOREM II.23 (Kisyński [76]). *Let* $\mathcal{H}_{+1} \subset \mathcal{H} \subset \mathcal{H}_{-1}$ *be a scale of spaces. Let* $H(t)$: $\mathcal{H}_{+1} \to \mathcal{H}_{-1}$ $(0 \le t \le 1)$ *be a family of* $\langle\ ,\ \rangle$ *(*$\mathcal{H}$ *inner product) self-adjoint operators, which are bounded (in* $\|\ \|_{-1,+1}$ *norm) and differentiable with a continuous derivative. Then, for any* $x_o \in \mathcal{H}_{+1}$, *there is a unique function* $x(t) \in \mathcal{H}_{+1}$ *which obeys:*

(i) $x(t)$ *is weakly* $\mathcal{H}_{+1}$*-continuous, i.e.* $\langle f,x(t)\rangle$ *is continuous for all* $f \in \mathcal{H}_{-1}$.

(ii) $\forall f \in \mathcal{H}_{+1}$, $\frac{d}{dt}\langle f,x(t)\rangle = -i\langle f,H(t)x(t)\rangle$.

(iii) $x(0) = x_o$.

This solution has $\|x(t)\| = \|x_o\|$ *and so* $U(t,s)$ *defined by* $U(t,s)x(s) = x(t)$ *extends to a unitary propagator on* $\mathcal{H}$.

Proof. See an appendix to this section for the case $H(t) = H_o + V(t)$. ∎ [21]

---

21 We only need $H(t)$ differentiable. With the extra hypothesis of C^1, Kisyński also proves $x(t)$ is $\|\ \|_{+1}$ continuous. Notice that $\|\ \|$-continuity of $x(t)$ follows trivially from the rest of the theorem.

THEOREM II.24 (Kisyński [76]). *Let* H(t) *obey all the conditions of Theorem II.23. Suppose, moreover, that* H(t) *is twice differentiable with a continuous second derivative. Then, whenever* $x_0 \in D(H(0))$, *the solution* x(t) *of* (i)-(iii) *has* $x(t) \in D(H(t))$ *and* $\frac{d}{dt} x(t)$ *exists in norm (and equals* $-iH(t)x(t)$).

Proof. See [76]. ∎

THEOREM II.25. *Let* $V(t) = V_1(t) + V_2(t)$ *be given,* $V_1(t) \in R$; $V_2(t) \in L^\infty$. *Suppose* $V_i(t)$ *has a bounded derivative (in the topology of* $R(L^\infty)$ *for* $i = 1(2)$). *Let* $H(t) = H_0 + V(t)$. *Then for any* $\psi_0 \in D(H_0^{1/2})$, *there is a solution of* $\frac{d}{dt} \langle\phi,\psi(t)\rangle = -i \langle\phi,H(t)\psi(t)\rangle$ *all* $\phi \in D(H_0^{1/2})$, $\psi(0) = \psi_0$. *The map* $\psi(s) \to \psi(t)$ *defines a unitary propagator. If, moreover,* $V_i(t)$ *is* C^2, *then for any* $\psi_0 \in D(H(0))$, $\psi(t)$ *as above has* $\psi(t) \in D(H(t))$ *and*

$$\frac{d}{dt} \psi(t) = -iH(t)\psi(t)$$

in the $\| \ \|$*-sense.*

Proof. Follows directly from Kisyński's results: Theorems II.23, II.24. ∎

As a consequence of this theorem:

COROLLARY II.26. *Let* $V \in R{+}L^\infty$. *Then, there is a unitary operator* U *with* $U[D(H_0)] = D(H)$; $U[Q(H_0)] = Q(H) \equiv Q(H_0)$.

Proof. Let $H(t) = H_0 + tV$. $U(1,0) = U$ obeys the theorem. ∎

Remark.

Since the high energy spectra of H and H_0 should be identical, this theorem is not surprising.

Appendix to II.7: A general theorem on time dependent evolution equations

THEOREM II.27. *Let* H_0 *be a positive operator on a Hilbert space and let* $\mathcal{H}_{+1} \subset \mathcal{H} \subset \mathcal{H}_{-1}$ *be its associated scale of spaces. Let* $H(t)$ *be a family of* $\mathcal{H}$*-symmetric operators* $(0 \leq t \leq 1)$ *from* $\mathcal{H}_{+1} \to \mathcal{H}_{-1}$ *so that for some* C *independent of* t:

(a) $C^{-1}(H_0 + 1) \leq H(t) \leq C(H_0 + 1)$.

(b) $B(t) = \frac{d}{dt}[H(t)^{-1}]$ *exists in the* $\| \ \|$*-sense and*

$$\|H(t)^{1/2}\, B(t)\, H(t)^{1/2}\| \leq C. \tag{II.20}$$

Then, for any $x_0 \in \mathcal{H}_{+1}$, *there is a unique function* $x(t) \in \mathcal{H}_{+1}$ *so that:*

(i) $x(t)$ *is continuous in the* $\mathcal{H}_{+1}$ *weak topology, i.e. for all* $\psi \in \mathcal{H}_{-1}$, $t \to \langle\psi, x(t)\rangle$ *is continuous.*

(ii) *For any* $f \in \mathcal{H}_{+1}$

$$\frac{d}{dt}\langle f, x(t)\rangle = -i\langle f, H(t)x(t)\rangle; \ x(0) = x_0.$$

Moreover:

(iii) $\lim\limits_{t \to t_0} \left\| \frac{x(t)-x(t_0)}{t-t_0} + iH(t)x(t) \right\|_{-1} = 0.$

(iv) $\|x(t)\| = \|x_0\|$.

(v) $x(t)$ *is* $\| \ \|$*-continuous.*

Thus, the map $U(t,s): x(s) \to x(t)$ *is unitary and extends to a unitary propagator.*

Remark.

As we mentioned in the text proper, this theorem (in a slightly different form) is contained in the work of Kisyński [76]. However, it is rather difficult to abstract just this result from Kisyński's extensive treatment.

Thus, for the reader's convenience, we present a proof which mimics Yoshida's proof of his theorem (Theorem II.21).[22]

Proof. The idea behind the proof (Yoshida's!) is quite elegant. One constructs approximate solutions $x_n(t)$ and shows that these are uniformly $\| \; \|_{-1}$ equicontinuous. From the weak compactness of Hilbert space balls, we find a weak limit $x(t) = \text{w-}\lim_{n' \to \infty} x_{n'}(t)$ for a subsequence. We show $x(t)$ is a solution obeying (i)-(iii) and that any solution of (i)-(iii) also obeys (iv) and (v) and is thus unique. Unicity of solutions obeying only (i) and (ii) follows from the existence of solutions obeying (i)-(iii).

(1) The approximate solutions

Let $H_n(t) = H(t)\,(1+n^{-1}H(t))^{-1}$, so that $\|H_n(t)\| \le n$. By (a), the $\| \; \|_{\pm 1}$ norms are equivalent to the $\langle u,(H(t)+1)^{\pm 1}u\rangle^{1/2}$ norms, so $H_n(t)$ is a bounded operator of $\mathcal{H}(\mathcal{H}_{\pm 1})$ into itself. As a result, we can uniquely solve the equation:

$$\frac{dz_n}{dt} = -iH_n(t)z_n(t);\; z_n(0) = z_0 \tag{II.21}$$

for $z_0 \in \mathcal{H}(\mathcal{H}_{\pm 1})$ and the solution will be in $\mathcal{H}(\mathcal{H}_{\pm 1})$. The solution is given by

$$z_n(t) = U_n(t,0)\, z_0$$

with $U_n(t,s)$ $(t > s)$ given by the familiar (and convergent!) time-ordered exponential:

$$U_n(t,s) = \sum_{m=0}^{\infty} (-i)^m \int_0^t ds_1 \int_0^{s_1} ds_2 \dots \int_0^{s_{m-1}} ds_m H_n(s_1)\dots H_n(s_m).$$

---

[22] Yoshida's techniques are also behind Kisyński's approach. He improves upon them and thereby obtains $\| \; \|_{+1}$ continuity by proving $\| \; \|_{+1}$ convergence of the $x_n(t)$ which we construct below. I discovered Kisyński's paper only after completing this proof myself and have not checked carefully the differences, if any, in the details of the two proofs.

Let $w_n(t) = \frac{d}{dt} z_n(t) = -iH_n(t)z_n(t)$ and let us compute

$$\frac{d}{dt} \langle w_n(t), H(t)^{-1} w_n(t)\rangle.$$

We first note that:

$$H_n(t)^{-1} = H(t)^{-1} + n^{-1}$$

so $\frac{d}{dt}(H_n{}^{-1}(t)) = B(t)$ and $\frac{d}{dt}(H_n(t)) = -H_n(t)B(t)H_n(t)$. We have:

$$\frac{d}{dt} w_n(t) = -iH_n(t)w_n(t) - H_n(t)B(t)w_n(t) \tag{II.22}$$

$$\frac{d}{dt}(H(t)^{-1} w_n(t)) = -iH^{-1}(t)H_n(t)w_n(t) - H^{-1}(t)H_n(t)B(t)w_n(t)+B(t)w_n(t).$$

Thus, using the self-adjointness of $H^{-1}(t)H_n(t)$, $\frac{d}{dt}\langle w_n(t), H^{-1}(t)w_n(t) =$ $-2\mathrm{Re}\ \langle w_n, H^{-1}H_nBw_n\rangle + \langle w_nBw_n\rangle$

$$\leq 2|\langle w_n(t), H^{-1}(t)H_n(t)B(t)w_n(t)\rangle| + \langle w_n(t), B(t)w_n(t)\rangle. \tag{II.23}$$

We now use the trick:

$$\begin{aligned}|\langle x, Ax\rangle| &= |\langle H(t)^{-1/2}x, H(t)^{1/2}A\, H(t)^{1/2}\, H(t)^{-1/2}x\rangle| \\ &\leq \|H(t)^{1/2}A\, H(t)^{1/2}\|\ \langle x, H(t)^{-1}x\rangle\end{aligned}$$

to obtain:

$$\frac{d}{dt}\langle w_n(t), H(t)^{-1}\, w_n(t)\rangle \leq 3C\ \langle w_n(t), H(t)^{-1}\, w_n(t)\rangle$$

where we have used (II.20) and the computation:

$$\begin{aligned}\|H(t)^{-1/2}\, H_n(t)\, B(t)H(t)^{1/2}\| &= \|(1 + n^{-1}H(t))^{-1}\, H(t)^{1/2}\, B(t)\, H(t)^{1/2}\| \\ &\leq \|H(t)^{1/2}\, B(t)\, H(t)^{1/2}\| \leq C.\end{aligned}$$

Thus, we conclude:

Whenever z_n *obeys (II.21) and* $w_n = \frac{dz_n}{dt}$, *we have* $(0 \leq t \leq 1)$:

$$\langle w_n(t), H(t)^{-1} w_n(t)\rangle \leq e^{3ct} \langle w_n(0), H(0)^{-1} w_n(0)\rangle . \tag{II.24}$$

Similarly:

$$\langle z_n(t), H^{-1}(t) z_n(t)\rangle \leq e^{ct} \langle z_o, H(0)^{-1} z_o\rangle \tag{II.25}$$

so $U_n(t,s)$ is uniformly bounded in t,s and n as a map from $\mathcal{H}_{-1}$ to $\mathcal{H}_{-1}$.

(2) Construction of x(t); props. (i) and (ii)

Let $x_0 \epsilon \mathcal{H}_{+1}$ and let $x_n(t)$ solve (II.21) with $x_n(0) = x_0$. Let $y_n(t) = \frac{d}{dt} x_n(t)$. Then (II.24) and (a) imply:

$$\begin{aligned} \|y_n(t)\|_{-1}^2 &\leq C \langle y_n(t), H(t)^{-1} y_n(t)\rangle \\ &\leq Ce^{3ct} \langle y_n(0), H(0)^{-1} y_n(0)\rangle \qquad \text{(II.26)} \\ &\leq Ce^{3ct} \langle x_o, (1+n^{-1}H(0))^{-1} H(0)\, (1+n^{-1}H(0)^{-1} x_o\rangle \\ &\leq Ce^{3ct} \langle x_o, H(0) x_o\rangle \leq C^2 e^{3ct} \|x_o\|_{+1}^2 \end{aligned}$$

i.e. $\|y_n(t)\|_{-1} \leq D \|x_0\|_{+1}$ (II.27)

where D is independent of n and $0 \leq t \leq 1$.

Thus the functions:

$$x_n(t) = x_0 + \int_0^t y_n(s)\, ds \tag{II.28}$$

are uniformly equicontinuous in the $\| \ \|_{-1}$ topology and are $\| \ \|_{-1}$ bounded. The balls in $\| \ \|_{-1}$ are sequentially compact in the weak $\mathcal{H}_{-1}$ topology so we can find a subsequence $x_{n'}(t) \to x(t)$ (weak $\mathcal{H}_{-1}$ topology).[23]

---

[23] Pick successive sub-subsequences convergent at the first rational, the first two rationals, etc. The ''diagonal'' of this sequence of sequences will converge at all rationals. By the equi-continuity, this will converge at all t's.

Since $\text{s-}\lim_{n\to\infty} (1+n^{-1}H(t))^{-1} = 1$ (in $\|\ \|_{+1}$ topology), we have for any $f \in \mathcal{H}_{+1}$:

$$\langle y_{n'}(t), H(t)^{-1}f\rangle = \langle -iH(t)\,(1+n'^{-1}H(t))^{-1}\, x_{n'}(t), H^{-1}(t)f\rangle$$

$$= \langle -ix_{n'}(t), (1+n'^{-1}H(t)^{-1}f\rangle \to \langle x(t),f\rangle .$$

Thus, for any fixed t and any $g = H(t)^{-1}f$ ($f \in \mathcal{H}_{+1}$), $\langle y_{n'}(t),g\rangle$ is Cauchy in n'. Since $\{H(t)^{-1}f \mid f \in \mathcal{H}_{+1}\}$ is dense in $\mathcal{H}_{+1}$ and the $y_{n'}(t)$ are bounded, we conclude $y_{n'}(t) \to y(t)$ (weak $\mathcal{H}_{-1}$) for some $y(t)$. Now, for any $f \in \mathcal{H}_{+1}$, we have

$$|\langle x(t),f\rangle| = |\langle y(t),H^{-1}(t)f\rangle$$
$$\leq \|y(t)\|_{-1} |H^{-1}(t)f\|_{+1}$$
$$\leq C\|y(t)\|_{-1}\|f\|_{-1} \tag{II.29}$$

so that $x(t) \in (\mathcal{H}_{-1})^{*}$, i.e., $x(t) \in \mathcal{H}_{+1}$ and $y(t) = -iH(t)x(t)$. From the weak convergence of $x_{n'}(t) \to x(t)$ and the $\|\ \|_{-1}$ equicontinuity of the x_n's, we conclude $x(t)$ is $\|\ \|_{-1}$ continuous. Thus $\langle f,x(t)\rangle$ is continuous for f in $\mathcal{H}_{+1}$. But (II.29) implies $x(t)$ is $\|\ \|_{+1}$ bounded so $\langle f,x(t)\rangle$ is continuous for any f in $\mathcal{H}_{-1}$ ($\mathcal{H}_{+1}$ is $\|\ \|_{-1}$ dense in $\mathcal{H}_{-1}$) i.e., (i) holds. Moreover, (II.28) implies:

$$\langle f,x_n(t)\rangle = \langle f,x_o\rangle + \int_o^t \langle f,y_n(s)\rangle\, ds$$

for any $f \in \mathcal{H}_{+1}$. Taking limits (and using the uniform boundedness of $\|y_n\|_{-1}$ to implement the dominated convergence theorem), we see:

$$\langle f,x(t)\rangle = \langle f,x_o\rangle + \int_o^t \langle f,y(s)\rangle\, ds = \langle f,x_o\rangle - i\int_o^t \langle f,H(s)x(s)\rangle\, ds. \tag{II.30}$$

From the weak $\mathcal{H}_{+1}$ continuity of x(t) and the $\| \ \|_{+1,-1}$ continuity of H(t), we conclude the integrand in (II.30) is continuous. Thus (ii) holds.

(3) Property (iii)

Since

$$\left|\left\langle f, \frac{x(t)-x(t_o)}{t-t_o} + iH(t)x(t)\right\rangle\right| = \left|\int_{t_o}^{t} \langle f, y(s)-y(t_o)\, ds\rangle\right| (t-t_o)^{-1}$$

$$\leq \|f\|_{+1} \sup_{t\leq s\leq t_o} \|y(s)-y(t_o)\|_{-1}$$

it is sufficient to prove the y(t) are $\| \ \|_{-1}$-continuous. This in turn will follow if we can show the $y_n(t)$ are $\| \ \|_{-1}$-continuous uniformly in n. Now the general equation

$$\frac{dq(t)}{dt} = -iH_n(t)\, q(t) + f(t);\ q(0) = q_o$$

is solved by:

$$q(t) = U_n(t,0)q_o + \int_0^t U_n(t,s)\, f(s)ds$$

so (II.22) implies

$$y_n(t) = U_n(t,0)\, y_n(0) - \int_0^t U_n(t,s)\, H_n(s)B(s)y_n(s)ds. \tag{II.31}$$

We have

$$\|H_n(s)B(s)y_n(s)\|_{-1} < C\|H(s)^{-\frac{1}{2}}\, H_n(s)B(s)H(s)^{\frac{1}{2}}\, H(s)^{-\frac{1}{2}}\, y_n(s)\|$$

$$\leq C^2\|H(s)^{-\frac{1}{2}}\, H_n(s)B(s)H(s)^{\frac{1}{2}}\|\, \|y_n(s)\|_{-1}$$

$$\leq C^3\|y_n(s)\|_{-1}$$

so $\|H_n(s)B(s)y_n(s)\|_{-1}$ is $\| \ \|_{-1}$ bounded uniformly in n and s. Since (II.25) tells us $U_n(t,s)$ is uniformly bounded from $\mathcal{H}_{-1}$ to $\mathcal{H}_{-1}$, we see that the integrand in the second term of (II.31) is $\| \|_{-1}$ bounded uniformly (in t, s, n) and thus this second term is continuous uniformly in n and t.

To complete the proof of (iii) we need only show $U_n(t,0)y_n(0)$ is uniformly (in n) continuous. First consider a special x_0 of the form $x_0 = H(0)^{-1}f$; $f \epsilon \mathcal{H}_{+1}$. Let $z_n(t) = U(t,0)y_n(0)$ so $\frac{dz_n}{dt} = iH_n(t)z_n(t)$; $z_n(0) = y_n(0)$. By (II.24):

$$\left\| \frac{dz_n}{dt} \right\|_{-1}^2 \leq Ce^{3ct} \langle H_n(0)y_n(0), H(0)^{-1}H_n(0)y_n(0)\rangle \leq C^2e^{3ct}\|f\|_{+1}^2.$$

Thus, when x_0 is of the form $H(0)^{-1}f$ ($f\epsilon\mathcal{H}_{+1}$), $U_n(t,0)y_n(0)$ is continuous uniformly in t and n.

Finally consider an arbitrary x_0. $\{H(0)^{-1}f \mid f\epsilon\mathcal{H}_{+1}\}$ is $\| \|_{+1}$ dense in $\mathcal{H}_{+1}$ so choose $x_0^{(m)} \to x_0$ (in $\| \|_{+1}$). The $H_n(0)$ are uniformly bounded as maps of $\mathcal{H}_{+1} \to \mathcal{H}_{-1}$ so $\|y_n(0) - y_n^{(m)}(0)\|_{-1} \to 0$ uniformly in n. Since the $U_n(t,0)$ are uniformly bounded as maps of $\mathcal{H}_{-1} \to \mathcal{H}_{-1}$ [(II.25)], $U_n(t,0)y_n^{(m)}(0) \to U_n(t,0)y_n(0)$ uniformly in t (with the convergence being uniform in n). Thus the $U_n(t,0)y_n(0)$ are equicontinuous so (iii) is proven.

(4) Properties (iv) and (v)

We have:

$(t-t_0)^{-1} [\langle x(t),x(t)\rangle - \langle x(t_o),x(t_o)\rangle] = (I)+(II)+(III)+(IV)$ where

$(I) = \langle x(t)-x(t_o), (t-t_o)^{-1} [x(t)-x(t_o)]-y(t_o)\rangle \to 0$ since $\|x(t)-x(t_0)\|_{+1}$ is bounded and (iii) holds.

$(II) = \langle x(t)-x(t_o), y(t_o)\rangle \to 0$ by the $\mathcal{H}_{+1}$ weak continuity of x.

$(III) = \langle x(t_o), (t-t_o)^{-1} [x(t)-x(t_o)]\rangle \to \langle x(t_o)-iH(t_o)x(t_o)\rangle$.

$(IV) = \langle (t-t_0)^{-1} [x(t)-x(t_0)], x(t_0)\rangle \to \langle -iH(t_0)x(t_0), x(t_0)\rangle$.

(III) and (IV) cancel, so $\frac{d}{dt}\|x(t)\|^2 = 0$, i.e. (iv) holds. (v) follows from the combined facts $x(t) \to x(t_o)$ (in $\mathcal{H}$-weak) and $\|x(t)\| \to \|x(t_o)\|$.

(5) Unicity

Let x(t) obey (i), (ii). Let f(t) solve the differential equation with $f(0) = f_o$ and with f obeying (iii). Then by a computation identical to (I) – (IV) above:[24]

$$\frac{d}{dt}\langle x(t), f(t)\rangle = 0.$$

Thus

$$\begin{aligned}\|x(t)\| &= \sup_{f\in\mathcal{H}_{+1},\ \|f\| = 1} |\langle x(t), f\rangle| \\ &= \sup_{f\in\mathcal{H}_{+1},\ \|f\| = 1} |\langle x(t), f(t)\rangle| \\ &= \sup_{f\in\mathcal{H}_{+1},\ \|f\| = 1} |\langle x(0), f(0)\rangle| = \|x(0)\|.\end{aligned}$$

As a result, the only solution obeying (i) and (ii) with $x_o = 0$ is $x(t) = 0$, i.e. unicity holds. ∎

Remarks

1. The unicity implies $x_n(t) \underset{W}{\to} x(t)$ since every subsequence has a convergent sub-subsequence.

2. In the Yoshida case (where (b) is replaced by $\|H(t)B(t)\| \leq C$) when we have H self-adjoint in addition, the proof is much simplified. A direct computation proves unitarity so (iii) is unnecessary. An argument analogous to step (1), only simpler, shows

---

24 In (I) we use the fact that the second factor obeys (iii) but we never need the fact that the first factor obeys (iii).

$$\frac{d}{dt}\|y_n(t)\|^2 = \langle y_n(t), H_n(t)B_n(t)y_n(t)\rangle \leq C\|y_n(t)\|^2.$$

After the analogue of step (2) the proof is complete.

3. (a) implies that $\|H(t)^{1/2}(H_0+1)^{-1/2}\| \leq C$ and $\|(H_0+1)^{-1/2}H(t)^{-1/2}\| \leq C$ so that under condition (a), (II.20) is equivalent to

$$\|(H_0+1)^{1/2}B(t)(H_0+1)^{1/2}\| \leq D$$

for D independent of t.

4. In the second edition of his book [T29, pp. 424-431], Yoshida presents a discussion of evolution equations based on a different approximation procedure. One approximates H(t) by piecewise continuous H's which are solvable by piecing together Stone's theorem solutions. One then shows these solutions converge in $\|\ \|$. Such an approach might be convenient for a discussion of time-dependent Feynman path integrals. (See also Faris [30].)

COROLLARY II.28. *Let* H_0 *be a positive operator on a Hilbert space. Let* V(t) *be a family of self-adjoint operators with the following properties:*

(a′) $Q(V(t)) \supset Q(H_0)$ *and* $a < 1$, b *independent of* t *so*

$$|\langle\psi, V(t)\psi\rangle| \leq a\langle\psi H_0\psi\rangle + b\langle\psi\psi\rangle$$

of all $\psi \epsilon Q(H_0) \equiv \mathcal{H}_{+1}$.

(b′) *There exists a derivative* $\dot{V}(t)$ *in the sense that*

$$\frac{d}{dt}[(H_0+1)^{-1/2}V(t)(H_0+1)^{-1/2}] \equiv (H_0+1)^{-1/2}\dot{V}(t)(H_0+1)^{-1/2}$$

and $|\langle\psi, \dot{V}(t)\psi\rangle| \leq a\langle\psi, H_0\psi\rangle + b\langle\psi, \psi\rangle$. *Then* (a) *and* (b) *of Theorem II.27 hold for* $\tilde{H}(t) = H_0 + V(t) + E$ (*some* E *fixed*) *and so the conclusions of Theorem II.28 hold with* $H(t) = H_0 + V(t)$.

Proof: By (a′), all the H(t)'s are bounded below uniformly in t, so we can find E with $\tilde{H}(t) = H_o + V(t) + E \geq 1$. Then by our general scale arguments, (a) holds for $\tilde{H}$. Since $0 \notin \text{spec}(\tilde{H})$, we have that:

$$(\tilde{H}(t))^{-1} = (H_o + 1)^{-\frac{1}{2}} [1 + (H_o + 1)^{-\frac{1}{2}} (V(t) + E)(H_o + 1)^{-\frac{1}{2}}]^{-1} (H_o + 1)^{-\frac{1}{2}}.$$

By (b′), $Q(t) = \frac{d}{dt} [(H_o + 1)^{-\frac{1}{2}} \dot{V}(t)(H_o + 1)^{-\frac{1}{2}}]$ exists and is bounded.

Thus

$$(H_o + 1)^{\frac{1}{2}} \frac{d}{dt} [(\tilde{H}(t))^{-1}] (H_o + 1)^{\frac{1}{2}} = [1+(H_o+1)^{-\frac{1}{2}} (V(t)+E)(H_o+1)^{-\frac{1}{2}}]^{-1}$$

$$Q(t) [1+(H_o+1)^{-\frac{1}{2}} (V(t)+E)(H_o+1^{-\frac{1}{2}}]^{-1}$$

exists and is bounded, so (b) holds. ∎

Remarks

1. If $A: \mathcal{H}_{+1} \to \mathcal{H}_{-1}$, then

$$\|A\psi\|_{-1} = \|(H_o + 1)^{-\frac{1}{2}} A(H_o + 1)^{-\frac{1}{2}}\| \, \|(H_o + 1)^{\frac{1}{2}}\psi\|$$

so $\|A\|_{-1,+1} = \|(H_o + 1)^{-\frac{1}{2}} A(H_o + 1)^{-\frac{1}{2}}\|$. Thus (b) says that $H(t): \mathcal{H}_{+1} \to \mathcal{H}_{-1}$ is differentiable as a map of $\mathcal{H}_{+1} \to \mathcal{H}_{-1}$ and has a bounded derivative (in $\| \ \|_{-1,+1}$ norm). This is Kisyński's theorem.

2. The condition (b′) that $|\langle u, V(t)u\rangle| \leq a\langle u, H_o u\rangle + b\langle u,u\rangle$ is very natural; for formally speaking, a solution, u(t), of the Schrödinger equation obeys:

$$\frac{d}{dt} \langle u, H(t)u\rangle = \langle u, \dot{V}(t)u\rangle.$$

Thus (b) is a condition on the growth of the energy of a state in time. In fact, it assures no worse than an exponential growth.

II.8. *Analyticity in the Coupling Constant*

We recall the following definition (see [T17], pp. 366-367):

DEFINITION. A family of closed operators, $T(\kappa)$, depending on a complex parameter κ, is said to be holomorphic in κ at $\kappa = 0$, if and only if, for some $\lambda \notin \operatorname{spec}(T(0))$; $(T(\kappa) - \lambda)^{-1}$ exists and is a bounded operator-valued analytic function for $|\kappa|$ sufficiently small.

If one such λ exists, the result is true for all λ not in spec $T(0)$.

When $T(\kappa)$ is holomorphic, one can show isolated non-degenerate eigenvalues are analytic in κ ([T17], pp. 368-371). At points of degeneracy where $T(\kappa)$ is self-adjoint on a line through the point (e.g. κ real with $T(\kappa)^* = T(\bar{\kappa})$), there is still analyticity ([T17], pp. 385-387) in κ.

One can ask when $H_0 + \kappa V$ will be holomorphic in κ for $V \epsilon L^\infty + R$. There is a completely satisfactory answer:

THEOREM II.29. *Let* $V \epsilon L^\infty + R$. *Then* $H_0 + \kappa V$ *is a holomorphic family of type (B) ([T17], p. 395) in the entire* κ *plane, and thus it is* holomorphic *in the above sense. In particular, the discrete (i.e. negative energy) eigenvalues are analytic in the coupling constant.*

Proof. This is a direct consequence of Theorem VII. 4.8 in Kato's book ([T17], p. 398). Alternately, see the appendix to this section. ∎

We also remark that as a result of a theorem of Rellich [97], (see also [T17], p. 402), we have:

THEOREM II.30. *Let* $D(H_0) \cap D(V)$ *be a core for* $H_0^{1/2}$ *with* $V \epsilon L^\infty + R$. *Then* $H_\kappa = H_0 + \kappa V$ *defined by the forms method is the unique self-adjoint extension of* $H_0 + \kappa V$ *defined on* $D(H_0) \cap D(V)$ *which is holomorphic at* $\kappa = 0$.

Proof. A direct result of Theorem VII. 4.13 of Kato's book. ∎

Appendix 1 to II.8. Analyticity and Compactness

The analytic perturbation theory of Rellich and Kato [97, T17] is based on the analyticity of the resolvent $R(\kappa,\lambda) = (T(\kappa) - \lambda)^{-1}$ and of the "spectral projections" $P_C(\kappa) = (2\pi i)^{-1} \oint_C R(\kappa,\lambda)d\lambda$ where C is a curve in the complex plane. Holomorphic families of type (A) and (B) [which yield analytic resolvents] are *essentially* defined by demanding that

$$F_E = V(H_0 - E)^{-1} \qquad \text{(II.32)}$$

or respectively

$$S_E = (H_0 - E)^{-1/2}\, V(H_0 - E)^{-1/2} \qquad \text{(II.33)}$$

be bounded and have norm strictly less than 1 as $E \to -\infty$. It is a striking feature of quantum mechanics in *three dimensional space* that F_E (or S_E) is compact in almost all cases of interest. Under this assumption, the results of perturbation theory follow from the analytic Fredholm theory (Theorem A.27). While this theory is less conceptual than the Kato-Rellich theory, it is simpler to derive from first principles.

Example 1. $V \epsilon L^2 + (L^\infty)_\varepsilon$; $H_0 = -\Delta$.

Write $V = V_{1,\varepsilon} + V_{2,\varepsilon}$; $V_{1,\varepsilon} \epsilon L^2$; $\|V_{2,\varepsilon}\| \leq \varepsilon$. Then $V_{1\varepsilon}(H_0 - E)^{-1}$ is Hilbert-Schmidt and $\|V_{2,\varepsilon}(H_0 - E)^{-1}\| < \varepsilon|E|^{-1}$ so $F_E = V(H_0 - E)^{-1}$ is compact.

Example 2. $H_0 = -\Delta + r^2$; $V \underset{\text{T.K.}}{<<} H_0^\alpha$ some $\alpha < 1$.

Then $F_E = [V(H_0 - E)^{-\alpha}]\,(H_0 - E)^{-1+\alpha}$ is compact since the first factor is bounded and the second factor is compact.

Example 3. $V \epsilon R + (L^\infty)$; $H_0 = -\Delta$.

By an argument, as in Example 1, S_E is compact.

We first remark that the use of:

$$(H_0 + \kappa V - E)^{-1} = (H_0 - E)^{-1} (1 + \kappa F_E)^{-1} \qquad \text{(II.34)}$$

or

$$(H_0 + \kappa V - E)^{-1} = (H_0 - E)^{-1/2} (1 + \kappa S_E)^{-1} (H_0 - E)^{-1/2} \qquad \text{(II.35)}$$

allows us to prove analyticity of the resolvents directly from compactness and so one can "return" directly to the Kato-Rellich approach.[25] One can also prove the analyticity of eigenvalues directly:

THEOREM II.31. *Let* H_0 *be a positive self-adjoint operator and let* S_E (*or* F_E) *be defined in the canonically cut plane and suppose it is compact there. Suppose also* $\lim_{E \to -\infty} \|S_E\| = 0$. *Then any isolated non-degenerate eigenvalues of* $H_0 + \kappa_0 V$ *are analytic in* κ *near* κ_0. *An n-fold degenerate level is at worst an* n^{th} *order branch point of a function analytic in a many sheeted neighborhood of* κ_0. *If* V *is symmetric and* κ_0 *is real, then even in this latter case, the eigenvalues are analytic at* κ_0.

Proof. By the analytic Fredholm theory, $(1 + \kappa S_E)^{-1}$ for any fixed κ is meromorphic in the canonically cut E plane. Moreover, the position of the poles is given locally[26] as the zeros of a holomorphic function $f(E,\kappa)$

---

[25] We note that (II.34) and (II.35) are essentially behind the Kato proofs of analyticity of the resolvent for families of type (A) and (B).

[26] By this we mean: If $(1 + \kappa_0 S_E)^{-1}$ has a pole at $E = E_0$, then for κ near κ_0 all the poles of $(1 - \kappa S_E)^{-1}$ with E near E_0, are given as the zeros ...

of two variables.[27] Thus, the theorem follows except for the final statement. This last assertion is a consequence of the fact that an analytic function with an nth order branch point on the real axis, real on the real axis *on every sheet* has $n = 1$. ∎

Appendix 2 to II.8. Continuity and the Min-Max Principle

For our treatment in later chapters, the full analyticity of the resolvent is not needed; all we ever use is continuity of the eigenvalues. In this appendix, we wish to point out a simple proof of the continuity based on the Weyl min-max principle:

PROPOSITION II.32. Let H be a self-adjoint operator which is bounded below. Let:

$$\mu_n(H) = \max_{\Psi_1,\dots,\Psi_{n-1}} \left[\min_{\Phi \epsilon Q(H)\ [\Psi_1,\dots,\Psi_{n-1}]^\perp;\ |\Phi| = 1} (\Phi, H\Phi)\right].$$

Then, either:

(a) $\mu_n(H)$ is the nth eigenvalue for H counting multiplicity.

or

(b) $\mu_n(H)$ is the bottom of the essential spectrum in which case $\mu_n = \mu_{n+1} = \dots$ and there at most $n-1$ eigenvalues.

Proof. A standard exercise in spectral analysis. ∎

THEOREM II.33. *The eigenvalues of* $H_0 + \kappa V$ $(V \epsilon R + (L^\infty)_\varepsilon;\ \kappa$ real) *are continuous in* κ.

Proof. Define, for each $\Psi \epsilon Q(H_0)$ with $\|\Psi\| = 1$:

$$f_\Psi(\kappa) = \min(\langle\Psi, H_\kappa \Psi\rangle, 0).$$

---

27 In the proof of the analytic Fredholm theorem (see e.g. [54]), the poles arise as the roots of a determinant whose elements are analytic in all variables of analyticity for κS_E. The multiplicity of the zero is the degeneracy of the eigenvalue.

The $f_\Psi(\kappa)$ are continuous. If we can show they are equicontinuous, then the μ_n's will be continuous and the proof will be complete. Given $\kappa_0 > 0$, pick b so that:

$$|\langle\Psi,V\Psi\rangle| \le (2\kappa_0 + 2)^{-1} \langle\Psi,H_0\Psi\rangle + b\langle\Psi,\Psi\rangle \qquad \text{(II.36)}$$

all $\Psi \in \mathcal{H}_{+1}$. Suppose $f_\Psi(\kappa) < 0$ for some $\kappa < \kappa_0 + 1$. Then

$$\langle\Psi, H_0 + \kappa V, \Psi\rangle < 0$$

which implies:

$$\langle\Psi, H_0\Psi\rangle \le -\kappa\langle\Psi, V\Psi\rangle \le \kappa|\langle\Psi,V\Psi\rangle| \qquad \text{(II.37)}$$

Thus:

$$|\langle\Psi,V\Psi\rangle| \le \tfrac{1}{2}\kappa\,(\kappa_0 + 1)^{-1}\, |\langle\Psi,V\Psi\rangle| + b\langle\Psi,\Psi\rangle \qquad \text{[by (II.36 and (II.37)]}$$

$$\le \tfrac{1}{2}\, |\langle\Psi,V\Psi\rangle| + b\langle\Psi,\Psi\rangle$$

so that:

$$|\langle\Psi,V\Psi\rangle| \le 2b \text{ if } f_\Psi(\kappa) < 0 \text{ for some } \kappa < \kappa_0 + 1.$$

Given , let $\delta = \min(1, (2b)^{-1}\varepsilon)$. If $|\kappa-\kappa_0| < \delta$, either $f_\Psi(\kappa) = f_\Psi(\kappa_0) = 0$ or $|\langle\Psi,V\Psi\rangle| \le 2b$. In the latter case:

$$|f_\Psi(\kappa) - f_\Psi(\kappa_0)| \le |\kappa - \kappa_0|\, |\langle\Psi V \Psi\rangle| \le \varepsilon. \blacksquare$$

II.9. *The Green's Function as an Integral Operator*

One is used to thinking of the resolvent $(H-E)^{-1}$ as having the form of a simple x-space integral operator for $E \notin \text{spec}(H)$.[28] In this section, we investigate this question for $V \in L^\infty + R$. This study is not motivated

---

28 In this case, we call its kernel the Green's Function.

by mere curiosity; it will play an essential role in studying scattering. In the L^2 case, Ikebe [56] has studied the Green's function and used it as the basis for proving asymptotic completeness (under additional assumptions).

When $V \epsilon R + L^\infty$, $V_{\|}^{1/2}(H_o - E)^{-1}$ is a bounded operator for E in the canonically cut plane (for $V_{\|}^{1/2} \underset{\text{T.K.}}{<<} H_o^{1/2} \underset{\text{T.K.}}{<<} H_o$). In this case, one can formally resum the Born expansion:

$$
\begin{aligned}
(H-E)^{-1} &= (H_o-E)^{-1} - (H_o-E)^{-1} V (H_o-E)^{-1} + \ldots \\
&= (H_o-E)^{-1} - [(H_o-E)^{-1} V^{1/2}] (1 - V_{\|}^{1/2} (H_o-E)^{-1} V^{1/2} + \ldots) [V_{\|}^{1/2} (H_o-E)^{-1}] \\
&= (H_o-E)^{-1} - [(H_o-E)^{-1} V^{1/2}] [1 + V_{\|}^{1/2} (H_o-E)^{-1} V^{1/2}]^{-1} [V_{\|}^{1/2} (H_o-E)^{-1}].
\end{aligned}
$$

Since the initial and final expression make sense as operators in $\mathcal{H}$, we expect the equality to be valid. In fact:

THEOREM II.34. *Let* $E \notin \operatorname{spec}(H) \cup \operatorname{spec}(H_o)$ *where* $H = H_o + V$ *with* $V \epsilon R + L^\infty$. *Then:*

$$
(H-E)^{-1} - (H_o-E)^{-1} = \tag{II.38}
$$

$$
- [(H_o-E)^{-1} V^{1/2}] [1 + V_{\|}^{1/2} (H_o-E)^{-1} V^{1/2}]^{-1} [V_{\|}^{1/2} (H_o-E)^{-1}].
$$

Proof. The technique is similar to that used in proving Tiktopoulos' formula. Let E be so negative that the Born expansion is convergent. Viewing $(H_o-E)^{-1}$:

$$
\mathcal{H}_{-1} \to \mathcal{H}_{+1};\ V_{\|}^{1/2}: \mathcal{H}_{+1} \to \mathcal{H};\ V^{1/2}: \mathcal{H} \to \mathcal{H}_{-1},
$$

we see that all rearrangements are valid. In this final form, all operators make sense as maps of $\mathcal{H} \to \mathcal{H}$. To extend from very negative E to all admissible E, we merely note both sides are analytic in E in the admissible region. ∎

As a direct result of multiplying (II.38) on the left by $V_{||}^{1/2}$ and/or on the right by $V^{1/2}$, we conclude:

COROLLARY II.35. *$V_{||}^{1/2}(H-E)^{-1}V^{1/2}$ is Hilbert-Schmidt when $V \epsilon R$ and $E \notin$ spec H.* ∎

COROLLARY II.36. *$V_{||}^{1/2}(H-E)^{-1}$ and $(H-E)^{-1}V^{1/2}$ are Hilbert-Schmidt when $V \epsilon L^1 \cap R$ and $E \notin$ spec H.* ∎

(II.38) also implies the Green's function is an integral operator:

THEOREM II.37. *Let $E \notin$ spec (H) where $H = H_0 + V$ with $V \epsilon R \cap L^1$. Then:*

(a) $(H-E)^{-1} - (H_0 - E)^{-1}$ is trace class.

(b) $(H-E)^{-1}$ is an integral operator of Carleman type, i.e., $(H-E)^{-1}$ has a kernel $G(x,y;E)$ so that for y fixed $G(\cdot,y;E) \epsilon L^2(\cdot)$ and for x fixed $G(x,\cdot;E) \epsilon L^2(\cdot)$.

Proof. $E \notin$ spec (H_0) ($\subset$ spec (H), as we shall see). $V_{||}^{1/2}(H_0-E)^{-1}$ has kernel $(4\pi)^{-1}|V(x)|^{1/2}|x-y|^{-1}e^{+i\kappa|x-y|}$ where $\kappa^2 = E$, Im $\kappa > 0$. This kernel is Hilbert-Schmidt. Thus, by Theorem II.34, $(H-E)^{-1} - (H_0-E)^{-1}$ is the product of two $\mathcal{I}_2$-operators (see Definition A.29) and a bounded operator; therefore, it is trace class. In particular, it has a Hilbert-Schmidt kernel and thus a Carleman kernel. Since $(H_0-E)^{-1}$ has a Carleman kernel (which is not Hilbert-Schmidt), $(H-E)^{-1}$ is of Carleman type. ∎

Part (a) of Theorem II.37 which will play the more essential role in the sequel depends on $V \epsilon L^1$. However, (b) does not depend on $V \epsilon L^1$; in fact:

THEOREM II.38. *Let $E \notin$ spec (H) where $H = H_0 + V$ with $V \epsilon R$. Then:*

(a) $(H-E)^{-1} - (H_0-E)^{-1}$ is Hilbert-Schmidt.

(b) $(H-E)^{-1}$ is an integral operator of Carleman type.

Proof. Let $A = (H_0-E)^{-1/2}\, V(H_0-E)^{-1/2} \in \mathcal{I}_2$. Since

$$(1+A)^{-1} - 1 = A(1+A)^{-1},$$

we see $(1+A)^{-1} - 1 \in \mathcal{I}_2$ when $(1+A)^{-1}$ exists. According to the Tiktopoulos' formula, $(H-E)^{-1} - (H_0-E)^{-1} = (H_0-E)^{-1/2}\, [(1+A)^{-1}-1]\, (H_0-E)^{-1/2}$, so (a) follows immediately. As in Theorem II.37, (b) follows from (a). ∎

Remark

There is an analogous treatment of $V \in L^2$ based on

$$(H-E)^{-1} = (H_0-E)^{-1}\, [1+V(H_0-E)^{-1}]^{-1}$$

in place of the Tiktopoulos' formula. Such a treatment would simplify the proof of some of Ikebe's easier results [56].

As a last result of this type, we note:

THEOREM II.39. *If* $V \in R$, $V_{\|}^{1/2}\,(H-E)^{-1}$ *is an integral operator of Carleman type.*

By (II.38), $V_{\|}^{1/2}\, [(H-E)^{-1} - (H_0-E)^{-1}]$ is Hilbert-Schmidt and thus of Carleman type, we need only show $V_{\|}^{1/2}\,(H_0-E)^{-1}$, which has kernel $\kappa(x,y) = |V(x)|^{1/2}\,(4\pi)^{-1}\, e^{i\kappa|x-y|}\, |x-y|^{-1}$, is of Carleman type. For fixed x, it is clearly L^2 in y. Pick $f \in R$ with $f(y) > 0$, all y. Then $\int |\kappa(x,y)|^2\, |f(y)|\, d^3x d^3y < \infty$, so $\int |\kappa(x,y)|^2\, |f(y)|\, d^3x < \infty$ a.e. in y. Thus $\kappa(\cdot,y) \in L^2(\cdot)$ a.e. in y which completes the proof. ∎

As one expects, G obeys various integral equations:

THEOREM II.40. *Let* $V \in R$, $E \notin \mathrm{spec}\,(H)$. *Then:*

(a) G *satisfies the integral equation:*

$$G(x,y;E) = G_0(x,y;E) - \int G_0(x,y;E) V(z) G(z,y;E)\, dz \qquad \text{(II.39)}$$

where G_0 *is the free Green's function:*

$$G_0(x,y;E) = e^{i\kappa|x-y|}/4\pi|x-y| \quad (\kappa = \sqrt{E};\ \mathrm{Im}\,\kappa > 0).$$

(b) $V^{1/2}$ G *is of Carleman type and is the unique solution of* (II.39) *with this property.*

(c) G *is symmetric in* x *and* y.

(d) $\overline{G(x,y;E)} = G(x,y;\overline{E})$

Proof. (a) If we take (II.38) and multiply by $V_{\|}^{1/2}$, we see:

$$V_{\|}^{1/2}(H-E)^{-1} = V_{\|}^{1/2}(H_o-E)^{-1} - B[1+B]^{-1}[V_{\|}^{1/2}(H_o-E)^{-1}]$$

$$= [1 - B(1+B)^{-1}][V_{\|}^{1/2}(H_o-E)^{-1}]$$

$$= (1+B)^{-1}[V_{\|}^{1/2}(H_o-E)^{-1}]$$

where $B = V_{\|}^{1/2}(H_o-E)^{-1}V^{1/2}$. Thus, using (II.38) again, we conclude:

$$(H-E)^{-1} - (H_o-E)^{-1} = -[(H_o-E)^{-1}V^{1/2}][V_{\|}^{1/2}(H-E)^{-1}]$$

which implies (a).[29]

(b) By Theorem II.39, $V^{1/2}$ G is of Carleman type. To prove uniqueness, we appeal to Section III.2 where we will see that the kernel $R_\kappa(x,y) = -V^{1/2}(x)\, G_o(x,z;E)V^{1/2}(z)$ has no homogeneous solution when $E \notin \text{spec}(H)$. Thus

$$\psi(x) = V^{1/2}(x)\, G_o(x,y;E) + \int R_\kappa(x,z)\,\psi(z)\,dz$$

and $\psi \in L^2$ implies $\psi(x) = V^{1/2}(x)\, G(x,y,E)$. This proves the uniqueness of part of (b).

(c) Following Ikebe [56], p. 7, we introduce the indefinite "inner" product:

$$\langle f,g\rangle_{ind} = \int f(x)\, g(x)\, d^3x.$$

---

29 This could also be proven directly by a resummation procedure analogous to the proof of (II.38).

To show G is symmetric, we need only show

$$\langle f, (H-E)^{-1}g\rangle_{ind} = \langle (H-E)^{-1}f,g\rangle_{ind}.$$

Since $\langle\ ,\ \rangle_{ind}$ is $\|\ \|_2$-continuous, we can suppose $f,g \in \mathrm{Ran}(H-E)$. Then we need only show $\langle\psi,H\phi\rangle_{ind} = \langle H\psi,\phi\rangle_{ind}$; $\psi\ \phi \in D(H) \subset \mathcal{H}_{+1}$, and this is immediate.

(d) We have $[(H-E)^{-1}]^* = (H-E)^{-1}$ so $\overline{G(x,y;E)} = G(y,x;\bar{E})$. Thus (c) implies (d). ∎

As an important application of this theorem, we prove an addendum to Theorem II.37:

THEOREM II.41. *Let* $E \notin \mathrm{spec}\,(H)$, *where* $H = H_0 + V$ *with* $V \in R \cap L^1$. *Then Green's function* $G(x,y;E)$ *has the property* $G(\cdot,y;E) \in L^1(\cdot)$ *for almost every fixed* y.

Proof. By the integral equation, (II.39), we need only show

$$\int |G_0(x,z;E)|\ |V(z)|\ |G(z,y;E)|\ dzdx < \infty$$

a.e. in y. The x integration is finite independent of z, so we need only show $\int |V(z)|\ |G(z,y;E)|\ dz < \infty$. By Corollary II.36, $|V(z)|^{1/2} G(z,y;E) \in L^2(z)$ a.e. in y. Since $|V(z)|^{1/2} \in L^2$, the result is proven. ∎

Addendum to Chapter II: The Rollnik Condition in $R^n (n \neq 3)$.

First, let us consider $n > 3$. The essence of the Rollnik condition is that $\lim_{E\to 0} V_{\|}^{1/2} (E-H_0)^{-1} V^{1/2}$ is a Hilbert-Schmidt operator. The kernel for H_0^{-1} is $g(x-y)$ where g is the Fourier transform of κ^{-2}. On covariance grounds, this must be Cr^{n-2} $(n > 2)$ so the analogue of the Rollnik condition is:

$$\int \frac{|V(x)|\ |V(y)|}{|x-y|^{2n-4}}\ d^n x\ d^n y < \infty\ (n > 3). \qquad \text{(II.40)}$$

If $n \geq 4$; $\int d^n x \, |x-y|^{-[2n-4]}$ is divergent near $x = y$ so *there are essentially no Rollnik potentials in dimensions bigger than 3.*[30]

This strange result is due to our picking the Rollnik condition from the class of potentials with $|V|^{1/2} << H_0^{1/2}$. For example, in R^3, $V \in L^2$ is equivalent to $(E-H_0)^{-1}V$ Hilbert-Schmidt. In higher dimensions, this Hilbert-Schmidt condition is as bad as the Rollnik condition. However, the condition $V << H_0$ which is also (nearly) equivalent to $V \in L^2$ has an extension to $n > 3$. In fact, Nelson [91] has shown $V \in L^s (s > n/2)$ implies $V << H_0 (n > 3)$.

In dimensions smaller than 3, the situation is much improved. For example, in dimension 1, a local function V going to 0 at ∞, has $Q(V) \supset Q(H_0)$ if and only if $V \in L^1 + (L^\infty)_\varepsilon$ and in that case $V^{1/2} << H_0^{1/2}$. Actually, it is not hard to see that this analysis holds for any finite measure (and for other measures!), e.g. $V(x) = \delta(x)$. Thus, we can define a unique Hamiltonian H with $Q(H) = \{\hat{\psi}(x) \mid \int \kappa^2 \psi(\kappa) \, \kappa < \infty\}$ so that $\langle\psi, H\psi\rangle = \|\psi'\|^2 + |\psi(0)|^2$.[31]

30 A detailed examination of the existence of weird V's obeying the Rollnik condition seems fruitless. We note that if V is locally bounded away from zero at any point, (II.40) fails; in particular if V is continuous at a point x with $V(x) \neq 0$, then (II.40) is invalid.

31 Over $\mathbb{R}$, $Q(H_0)$ only contains bounded continuous functions, so $\psi(0)$ is well defined.

CHAPTER III

BOUND STATES

III.1. *Introduction*

In this chapter, we will study bound states for Hamiltonians of the type defined in Chapter II. Even if bound states were not important in and of themselves, some of the material of this chapter would be essential in scattering theory. For, as we have already seen, the kernel $R_k(x,y)$ defined in (I-17) arises naturally in several places. A central property of this kernel is that $(1-R_k)^{-1}$ exists [when $V \epsilon R + (L^\infty)_\varepsilon$] whenever E lies in the canonically cut plane and is not an eigenvalue of H, and this property can only be proven by a study of the bound states of H.

But bound states are of some physical interest and in particular, the following questions will be (or have been considered):

(1) Is the negative energy spectrum discrete? [1]

(2) Which of the various bounds on the number of eigenstates (e.g. Bargmann [4], Schwinger [101]) carry over to this case? (See Section III.3.)

(3) What can be said about the impossibility of positive energy bound states? (See Section III.4.)

(4) Do eigenfunctions obey the "usual" integral equations? (See Section III.2.)

(5) Do eigenfunctions have an exponential fall-off at ∞? (See appendix to Section VI.2.)

---

1 In Section II.3, we answered this question in the affirmative.

In the course of studying these questions, we will present some results which are new even in the usual $V \epsilon L^2 + L^\infty$ case; in particular

(a) When $V \epsilon R$, Schwinger's bound on the number of negative energy states, viz.:

$$N \leq \int \frac{|V(x)|\,|V(y)|}{|x-y|^2}\, d^3x\, d^3y$$

is actually a bound on the number of non-positive energy states, i.e. N can include zero energy bound states (and even zero-energy s-wave resonances).

(b) When $V \epsilon R$, there are no high-positive-energy bound states.[2]

(c) A simple proof that there are no positive energy bound states when

$$\int |x-y|^{-2}\, |V(x)|\, |V(y)|\, d^3x\, d^3y < (4\pi)^2$$

is presented.[3]

III.2. *The Integral Equation for Bound States*

In this section, we will derive an integral equation for negative energy solutions of the time independent Schrodinger equation ($H\psi = E\psi$). For comparison purposes, let us consider the situation when $V \epsilon L^2$. Then $H = H_0 + V$ is an operator equation, so that $H\psi = E\psi$ implies $(E - H_0)\psi = V\psi$ or:

$$\psi = [(E - H_0)^{-1} V]\psi\,. \tag{III.1}$$

Since $V \epsilon L^2$, the operator $(E - H_0)^{-1}V$ has a simple Hilbert-Schmidt kernel, and so (III.1) has the form ($\kappa = \sqrt{-E}$):

---

2 Actually, we suspect that if $V \epsilon R$, there are no positive energy bound states; however, almost all approaches to proving non-existence of positive energy bound states require V to be smooth. The weak result (b) needs no smoothness.

3 This result follows from a general theorem of Kato [70], whose proof is rather complex.

$$\psi(x) = -\int d^3y \, \frac{e^{-\kappa|x-y|}}{4\pi\,|x-y|} V(y)\,\psi(y)\,. \tag{III.2}$$

If $V \epsilon R$, the crucial relation $H\psi = H_0\psi + V\psi$ (as a relation between vectors in $\mathcal{H}$!) may not hold since $D(H) \cap D(H_0)$ may be $\{0\}$. Thus (III.1) and hence (III.2) will fail. Let us make some formal manipulations on (III.1). If $\phi = V_{\|}{}^{1/2}\psi$ and we multiply (III.1) by $V_{\|}{}^{1/2}$, we obtain

$$\phi = [V_{\|}{}^{1/2}\psi\,(E - H_0)^{-1}V^{1/2}]\,\phi \tag{III.3}$$

an integral equation first employed by Schwinger [101] in the $L^2 \cap L^1$ case. When $V \epsilon R$, (III.3) makes sense and so we expect that it should be provable for bound states. Analogously to (III.1), (III.3) has an integral form with a Hilbert-Schmidt kernel ($\kappa = \sqrt{-E}$):

$$\phi(x) = -\int d^3y \left[V_{\|}{}^{1/2}(x)\, \frac{e^{-\kappa|x-y|}}{4\pi\,|x-y|}\, V^{1/2}(y) \right] \phi(y)\,. \tag{III.4}$$

For negative energies, the proof of (III.3) for eigenfunctions depends on:

LEMMA III.1. *Let* $\Psi \epsilon D(H)$ *with* $H = H_0 + V$ *in the forms sense. Then for any* E *in the canonically cut plane:*

$$V_{\|}{}^{1/2}(E-H_0)^{-1}(E-H)\Psi = (1 - V_{\|}{}^{1/2}(E-H_0)^{-1}V^{1/2})\,(V_{\|}{}^{1/2}\Psi). \tag{III.5}$$

Proof. We use the (by now) familiar trick of picking E so negative that formal power series manipulations are legitimate and then continue analytically. Pick E so that

$$\|V_{\|}{}^{1/2}(E-H_0)^{-1}V^{1/2}\| < 1; \text{ then } (1 - V_{\|}{}^{1/2}(E-H_0)V^{1/2})^{-1}$$

exists. As in Theorem II.40(a):

$$V_{\|}{}^{1/2}(E-H)^{-1}\;[1 - V_{\|}{}^{1/2}(E-H_0)V^{1/2}]^{-1}\;[V_{\|}{}^{1/2}(E-H_0)^{-1}].$$

Thus:

$$[1 - V_{\|}^{1/2}(E-H_o)^{-1}V^{1/2}]\, V_{\|}^{1/2}(E-H)^{-1} = V_{\|}^{1/2}(E-H_o)^{-1}.$$

For $\Psi \in D(H)$ apply this last equation to $(E-H)\Psi$. Thus (III.5) holds for E very negative. Since both sides are analytic for E in the cut plane, (III.5) holds in the entire cut plane. ∎

As an immediate corollary of this lemma we have:

THEOREM III.2. *Let* $V \in R + L^\infty$. *Let* $E = -\kappa^2 < 0$. *If* $H\Psi = E\Psi$, *then* $\Phi = V_{\|}^{1/2}\Psi$ *obeys* (III.3). *If moreover* $V \in R$, *then* Φ *obeys the Fredholm integral equation (III.4).* ∎

Of crucial importance is the converse of this last theorem. This depends on:

LEMMA III.3. *Let* $\Psi \in \mathcal{H}_{+1}$ *and let* $V \in R + L^\infty$. *Then:*

$$[1 - (E - H_o)^{-1}V]\Psi = (E - H_o)^{-1}(E - H)\Psi \tag{III.6}$$

for any E *in the canonically cut plane.*[4]

Proof. $E - H = (E - H_o) - V$ as maps of $\mathcal{H}_{+1} \to \mathcal{H}_{-1}$. ∎

THEOREM III.4. *Let* $V \in R + L^\infty$. *Let* $\phi = [V_{\|}^{1/2}(E-H_o)^{-1}V^{1/2}]\phi$ *with* E *in the canonically cut plane. Then, there is a* ψ *with:*

(a) $\psi \in D(H)$.

(b) $H\psi = E\psi$.

(c) $\phi = V_{\|}^{1/2}\psi$.

Proof. Let $\psi = [(E - H_o)^{-1}V^{1/2}]\phi$. Since $V^{1/2} \ll H_o^{1/2}$, $(E - H_o)^{-1/2}V^{1/2}$ is a bounded operator, so that $\psi \in \operatorname{Ran}(E - H_o)^{-1/2} = D(H_o^{1/2}) \equiv \mathcal{H}_{+1}$.

---

4 As usual, $(E - H)$ is viewed as a map from $\mathcal{H}_{+1} \to \mathcal{H}_{-1}$ and $(E - H_o)^{-1}$ from $\mathcal{H}_{-1} \to \mathcal{H}_{+1}$.

Because ϕ obeys the integral equation, (c) is clear. Thus

$$\psi = [(E - H_o)^{-1} V^{1/2}] (V^{1/2}\psi) = [(E - H_o)^{-1} V]\psi.$$

Then, by (III.6), $(E - H_o)^{-1} (E - H) \Psi = 0$. Since $(E - H_o)^{-1}$ is a bijection of $\mathcal{H}_{-1} \to \mathcal{H}_{+1}$, we have $(E - H) \Phi = 0$ as an element of $\mathcal{H}_{-1}$. But then $H\Psi = E\Psi \in \mathcal{H}_{+1} \subset \mathcal{H}$; i.e. $\Psi \in \mathcal{H}_{+1}$ and $H\Psi \in \mathcal{H}$, so $\Psi \in D(H)$. Thus (a) and (b) are true. ∎

COROLLARY III.5. *Let* $V \in R$.[5] *Then the integral equation*

$$\psi = \phi + [V_{\|}^{1/2} (E - H_o)^{-1} V^{1/2}]\psi$$

has a unique solution for all $\phi \in L^2$, *whenever* E *is a number in the canonically cut plane which is not an eigenvalue of* H; *in particular whenever* $\operatorname{Im} E \neq 0$.

Proof. Since $V_{\|}^{1/2} (E - H_o)^{-1} V^{1/2}$ is Hilbert-Schmidt, this is a direct consequence of the Fredholm alternative (Theorem A.25). ∎

When $E = 0$ or is positive, (III.1) no longer makes sense as an operator equation since $(E-H_o)^{-1} V$ cannot be defined. However (and this was first remarked by Scadron et al [100]), when $V \in R$, $V_{\|}^{1/2} (E-H_o)^{-1} V^{1/2}$ has a Hilbert-Schmidt limit as E approaches the positive real axis from *one* side (or the other). Thus, we expect an integral equation of type (III.3) to hold whenever we have a zero or positive energy eigenfunction. As we shall see, the analogue of Theorem III.2 is true, but the analogue of Theorem III.4 is not.

---

5 R_k is compact when $V \in R + (L^\infty)_\varepsilon$, so this corollary extends to that case.

THEOREM III.6. *Let* $H\psi = E\psi$ *for* $H = H_o + V$ *with* $V \epsilon R$ *and* $E \geq 0$. *Then:*

$$\lim_{\varepsilon \downarrow 0} (1 - V_{||}^{1/2} (E \pm i\varepsilon - H_o)^{-1} V^{1/2}) (V_{||}^{1/2}\psi) = 0,$$

i.e. $\phi = V_{||}^{1/2}\psi$ *obeys the pair of integral equations:*

$$\phi(x) = \int |V(x)|^{1/2} (4\pi|x-y|)^{-1} e^{\pm i\sqrt{E}|x-y|} V^{1/2}(y)\, \phi(y)\, d^3y. \qquad \text{(III.6}\pm\text{)}$$

Proof. By Lemma III.1,

$$V_{||}^{1/2}(E \pm i\varepsilon - H_o)^{-1} (E \pm i\varepsilon - H)\Psi = (1 - V_{||}^{1/2}(E \pm i\varepsilon - H_o)^{-1} V^{1/2}) (V_{||}^{1/2}\Psi) \qquad \text{(III.7)}$$

as 0, the right-hand side of (III.7) converges to $\eta_{\pm}$ where (III.6$\pm$) is equivalent to $\eta_{\pm} = 0$. Let χ_{Ω} be the characteristic function of a bounded set Ω. Then, using $(E - H)\Psi = 0$:

$$\chi_{\Omega}\eta_{\pm} = \lim_{\varepsilon \downarrow 0} \pm (i\varepsilon) (\chi_{\Omega} V_{||}^{1/2}) (E \pm i\varepsilon - H_o)^{-1}\Psi,$$

Since V is locally in L^1, $\chi_{\Omega} V_{||}^{1/2} (E \pm i\varepsilon - H_o)^{-1}$ is Hilbert-Schmidt with norm:

$$\left[\int \chi_{\Omega}(x)\, |V(x)|\, \frac{e^{-1\kappa|x-y|}}{4\pi|x-y|^2}\, d^3x\, d^3y \right]^{1/2} < \frac{1}{2\kappa^{1/2}} \left[\int_{\Omega} |V(x)|\, d^3x \right]^{1/2}$$

where $\kappa = |\mathrm{Im}\sqrt{E \pm i\varepsilon}| = 0(\varepsilon)$. Thus the term inside the last limit has norm $\sim \varepsilon 0(\varepsilon^{-1}) = 0(\varepsilon^{1/2}) \to 0$ as $\varepsilon \downarrow 0$. As a result $\chi_{\Omega}\eta_{\pm} = 0$ for any Ω, $\eta_{\pm} = 0$. ∎

Remark

It is not true however, that solutions of the integral equation (III.6$\pm$) necessarily correspond to bound states. For example, a spherical square-well with a zero-energy s-wave resonance has a wave function which satisfies (III.6±) (for $E = 0$) but which has $\psi(r) \quad r^{-1}$. Thus, $|V|^{1/2}\psi \in L^2$ but $\psi \notin L^2$, i.e. (III.6±) has an L^2-solution but $H\psi = 0$ has no L^2 solution.

III.3. *Bounds on the Number of Eigenvalues*

The problem of obtaining bounds on the number of eigenvalues from information on the potential was first considered by Bargmann [4]. Using the fact that the number of eigenvalues in a given ℓ-channel is equal to the number of zeros of the zero energy regular solution, he showed that $n_\ell(V) < (2\ell + 1)^{-1} \int r|V(r)|\, dr$. Schwinger [101] returned to this problem and recovered Bargmann's result by an integral equation method (see below). He also showed that the total number of *negative* energy eigenvalues is bounded by $\int |x-y|^{-2}\, |V(x)|\, |V(y)|\, d^3x\, d^3y$. It was in this paper that Schwinger independently rediscovered Rollnik's condition. Within the past few years, there has been a renewed interest in these bounds. Weinberg [116], Ghirardi and Rimini [42], Fonda and Ghirardi [34], have considered bounds on the total number of states while Calogero [12, 13, 14, T4], Frank [35] and Cohn [22, 23] have considered the number of states in a fixed ℓ-channel. There has been some interest in using these bounds to determine the asymptotic behavior of these quantities as the coupling constant goes to infinity, see e.g. Calogero [14], Chadan [19] and Simon [102].

In this section, we consider the question of carrying these bounds over to the case of potentials in R, etc. There are two general procedures for going about this. First, one can attempt a direct proof mimicking the $L^2 + L^\infty$ case. Not surprisingly, the Schwinger bound proof carries over directly; in fact, we will make a minor improvement in the result when $V \in R$. The second approach is to approximate V with Kato potentials and then use Corollary II.15 to obtain bounds for $N(V)$, etc. As an example of this method, we prove the bounds of Ghirardi-Rimini and Bargmann for our classes of potentials. The various other bounds and thereby the asymptotic behavior[6] can be carried over by similar means.

---

[6] Chadan's beautiful result cannot be carried over without some additional work which we have not done; i.e. only the gross asymptotic behavior of Calogero and Simon can be proven from the bounds alone.

Let us first consider Schwinger's bound. Given our preliminary spade work in Sections II.8 and III.2, the proof of his bound is identical to the one he gives in the $L^2 + L^\infty$ case:

LEMMA III.7 (Schwinger [101]). *Let* $N(V; E < -\kappa^2)$ *be the number of bound states of* $H_0 + V$ *with energy* E *bounded by* $-\kappa^2$. *Let* $V \epsilon R$ *and* $\kappa > 0$. *Then:*

$$N(V; E < -\kappa^2) \leq (4\pi)^{-2} \int d^3x\, d^3y\, |V(x)|\, e^{-2\kappa|x-y|}\, |x-y|^{-2}\, |V(y)|.$$

Proof. Consider first the case where $V \leq 0$. Following Schwinger, we note that by the continuity in coupling constant (Theorem II.33) and the fact that the eigenvalues decrease monotonically, $N(V; E < -\kappa^2)$ is the number of positive $\lambda < 1$ for which $(H_0 + \lambda V)\psi = -\kappa^2\psi$ has a solution. By Theorem III.2, this is identical to the number of solutions of

$$\phi(x) = \lambda \int d^3y\, R_\kappa(x,y)\, \phi(y) \tag{III.8}$$

with $\lambda < 1$, where R is given by (III.4). But[7]

$$\begin{aligned} \mathrm{Tr}\,(R_\kappa^+ R_\kappa) &= \sum_n (\lambda_n^{-2}) \\ &\geq \sum_{\lambda_n < 1} \lambda_n^{-2} \\ &\geq N(V; E < -\kappa^2) \end{aligned}$$

where λ_n is the set of all eigenvalues of (III.8). For V not obeying $V \leq 0$, we need only use:

$$N(V; E < -\kappa^2) \leq N(-|V|; E < -\kappa^2). \blacksquare$$

---

7 For $V < 0$, R_κ is self-adjoint and in $\mathcal{I}_2$.

THEOREM III.8 (Schwinger's Bound). *Let* $V \epsilon R$. *Then*

$$N(V;\, E \le 0) \le (4\pi)^{-2} \int d^3x\, d^3y\, |V(x)|\, |V(y)|\, |x-y|^{-2}.$$

Proof. From the lemma and

$$N(V;\, E < 0) = \lim_{\kappa \uparrow 0} N(V;\, E < -\kappa^2)$$

Schwinger concluded his bound. To include zero-energy states, we proceed differently. Suppose first $V \le 0$. Since R_κ is the kernel of $V^{1/2}(H_0 + \kappa^2)^{-1} V^{1/2}$, we see that $R_\kappa \le R_0$ (in the sense of operators), so (by the min-max principle) for each solution of $\phi = R_\kappa \phi$, there must be a solution of $\lambda^{-1}\psi = R_0\psi$ with $\lambda^{-1} > 1$. Moreover, by Theorem III.6, zero-energy bound states obey $\psi = R_0\psi$. Thus $N(V, E \le 0) \le \mathrm{Tr}\,(R_0^\dagger R_0)$. If $V \le 0$ does not hold, we use $N(V;\, E \le 0) \le N(-|V|;\, E \le 0)$.∎ [8]

Remarks

1. So far as we know, the inclusion of $E = 0$ states is a new result even for Kato V.

2. We can count zero-energy resonances which do not correspond to square-integrable eigenfunctions (e.g. s-wave resonances) in $N(V;\, E \le 0)$, for as we shall see in Section VI.2, such resonances are due to solutions of $R_0\psi = \psi$.

3. Of course $|V_{(-)}(x)| = |\min\,(V(x), 0)|$ can replace $V(x)$ in the integral of the inequality.

4. In the inequality, one can replace $\le$ with $<$ since R_κ must have eigenvalues with $\lambda > 1$.

As two examples of the approximation method, let us prove:

---

[8] To conclude this, one cannot merely use the min-max principle, Proposition II.32, because $E = 0$ is the bottom of the essential spectrum. But a quadratic form argument shows that $\dim P_{(-\infty,0]}\,(H_0 + V) \le \dim P_{(-\infty,0]}\,(H_0 + |V|)$ where $P_{(-\infty,0]}(A)$ is the non-positive energy spectral projection for A.

THEOREM III.9 (the Ghirardi-Rimini bound). *Let* $V \epsilon R$. *Then*

$$N(V; E < 0) \leq (4\pi)^{-2} \int d^3x\, d^3y\, V(x)\, V(y)\, |x-y|^{-2}.$$

Proof. Let $V_N(x) = \begin{cases} V(x) & \text{if } |V(x)| < N \\ 0 & |V(x)| > N \end{cases}$.

Then $V_N \to V$ in Rollnik norm and so by Corollary II.15,

$$P_N(-\infty, E) \to P(-\infty, E)$$

in $\| \ \|$ whenever $E < 0$ is not an eigenvalue of $H_0 + V$. Thus

$$N(V_N, E < -\kappa^2) \to N(V; E < -\kappa^2).$$

Since V_N is Kato, the Ghirardi-Rimini bound [42] tells us that

$$N(V_N; E < -\kappa^2) \leq N(V_N; E < 0) \leq (4\pi)^{-2} \int d^3x\, d^3y\, |x-y|^{-2}\, V_N(x)\, V_N(y).$$

By the dominated convergence theorem and $V \epsilon R$:

$$N(V; E < -\kappa^2) \leq (4\pi)^{-2} \int d^3x\, d^3y\, |x-y|^{-2}\, V(x)\, V(y)$$

which implies the result.∎

Remarks

1. As Ghirardi and Rimini note, this bound is always better than the $|V(x)|$ bound and is sometimes better, sometimes worse than the $|V_{(-)}(x)|$ bound.

2. There is also a direct proof of this theorem following Ghirardi and Rimini. For any solution of $(H - E)\psi = 0$ $(E < 0)$ has

$$[(H_0 - E)^{-1/2}\, V (H_0 - E)^{-1/2}]\phi = \phi$$

with $\phi = (H_0 - E)^{1/2}\psi$ (by an argument similar to those in Section III.2).

Since the binding energy is monotone in the coupling constant,[9] an argument identical to the Schwinger argument shows:

$$N \leq \lim_{E \downarrow 0} [\mathrm{Tr}\,[(H_o - E)^{-1/2} V (H_o - E)^{-1/2}]^2] = (4\pi)^{-2} \int d^3x\, d^3y\, V(x)\, V(y)\, |x-y|^{-2}.$$

THEOREM III.10 (Bargmann's bound). *Let* $V \in R$. *Suppose* V *is central and* $\int_o^\infty r|V(r)|\,dr < \infty$. *Then the number,* $n_\ell(V)$, *of bound states in the* ℓ*-channel obeys:*

$$n_\ell(V) \leq (2\ell+1)^{-1} \int_o^\infty r|V(r)|\,dr.$$

Proof. Let V_N be defined as in the proof of Theorem III.9. Let P_ℓ be the projection onto the subspace of angular momentum ℓ; P_ℓ commutes with $H_o + V$ and $H_o + V_N$ and their spectral projections. Thus

$$P_N(-\infty, E) P_\ell \xrightarrow{s} P(-\infty, E) P_\ell$$

if $E < 0$ is not an eigenvalue of $H_o + V$. Therefore:

$$\begin{aligned} n_\ell(V) &= \lim_{E \uparrow 0} \dim [P(-\infty, E)] \\ &= \lim_{E \uparrow 0} \lim_{N \to \infty} \dim [P_N(-\infty, E)] \leq (2\ell+1)^{-1} \int_o^\infty r|V(r)|\,dr \end{aligned}$$

since each $\dim P_N(-\infty, E) < (2\ell+1)^{-1} \int_o^\infty r|V_N(r)|\,dr$ by the usual Bargmann proof [4].∎

III.4. *Positive Energy Eigenvalues*

There is a very simple and appealing physical argument [18, pp. 30, 51] that assures us that a potential going to zero at ∞ cannot have bound

---

9 Ghirardi and Rimini give a complex argument for this when V is not purely negative, but one need only note $dE/d\lambda = \langle\psi|V|\psi\rangle = E(\lambda) - \langle\psi|H_o|\psi\rangle < 0$.

states (i.e. square integrable eigenfunctions) at positive energies: for $V \to 0$ and no potential wall can keep a quantum mechanical particle in since it can tunnel out. Nevertheless, Von Neumann and Wigner [93], and more recently Weidmann [115] have constructed explicit examples of potentials which possess positive energy eigenvalues! The physics behind this phenomena is quite simple: bumps in a potential will reflect a wave and so well-arranged bumps can act constructively and prevent a wave from reaching infinity, i.e., stationary waves can be formed with fall-off.

It is of interest to know when such eigenvalues do not occur – not only is such behavior surprising, it presents a formidable technical barrier in discussing certain aspects of scattering theory. Thus, Ikebe's proof [56] of asymptotic completeness in the two-body case depends on a result of Kato [69] which assures the non-occurrence of positive energy bound states when V is "regular" and $V(\vec{r}) = 0(r^{-1})$. Hepp's treatment of asymptotic completeness in the many body case [49] holds only for potentials without positive energy bound states[10] [and, in the many body case, there are very few results on the non-occurrence of such states; see, however, Weidmann [113, 114]].

In addition to Kato's result for $V = 0(r^{-1})$, Odeh [94] has shown that no positive energy eigenstates exist if V is strongly repulsive (i.e., $\frac{\partial V}{\partial r} < 0$ for r large) and smooth. These results have been subsumed by Simon [103]* who shows no positive energy states occur when V is smooth and $V = V_1 + V_2$; $V_1 = o(r^{-1})$; $V_2 \to 0$; $\lim_{r\to\infty} r\frac{\partial V_2}{\partial r} = 0$ (i.e., V_2 doesn't wiggle too much: for an exact statement see Theorem III.11 below). There are additional results in case V is central, see e.g., Weidmann [115].

---

10 Actually, the stronger statement that no bound states occur above the continuum is needed.

* Note added in proof: A similar result has been proven by S. Agmon, J. Anal. Math. 23 (1970), 1-25.

These results do not really depend on V being Kato; they merely rely on $\mathcal{D}_{\{\vec{r}|r>R\}} \subset D(H)$ for some R where $\mathcal{D}_{\{\vec{r}|r>R\}}$ is the set of C^∞ functions with compact support in $\{\vec{r}|r>R\}$. Thus we have:

THEOREM III.11. *Let* V *be a real-valued function on* R^3 *with the following properties:*

(a) $V \in R + L^\infty$ *and* $V = V_1 + V_2$[11] *with*

(b) *For some* R_0, V_1, V_2 *are* C^∞ *in* $\{\vec{r}|r>R_0\}$.

(c) $\lim_{r\to+\infty} r\, V_1(\vec{r}) = 0$

(d) $\lim_{r\to\infty} V_2(\vec{r}) = 0$

(e) $\lim_{r\to\infty} r \dfrac{\partial V_2}{\partial r}(\vec{r}) \equiv E_0 < \infty.$

Then $H_0 + V$ *has no eigenvalues in* (E_0, ∞).

Proof. The proof in [103] carries over without any essential change. ∎

There is another result on the absence of positive energy eigenvalues due to Kato. It differs from the others in that no smoothness is needed. Kato shows that when $\|V\|_R < 4\pi$; $H_0 + V$ and H_0 are unitarily equivalent. Thus, in this case, no positive eigenvalues can exist. Using Theorem III.6, we can recover Kato's result very simply:

THEOREM III.12. *Suppose* $\|V\|_R < 4\pi$. *Then* $H_0 + V$ *defined by the forms method has no positive energy bound states.*[12]

Proof. By Theorem III.6, if ψ is such a bound state, $\phi = V_{\|}^{1/2}\psi$ obeys $\phi = \kappa\phi$ where $\kappa(x,y)$ is a kernel with

$$\int |\kappa(x,y)|^2\, d^3x\, d^3y = (4\pi)^{-2}\|V\|_R^2 < 1$$

by assumption. Thus $\phi = \kappa\phi$ is impossible so the theorem is proven. ∎

---

[11] This breakup has nothing to do with the first one.

[12] By Schwinger's bound, it already has no non-positive energy eigenfunction.

We can also use Theorem III.6 to tell us something about positive energy bound states when V doesn't have $\|V\|_R < 4\pi$.

THEOREM III.13. *Let* $V \epsilon R$. *Then* $E_0 > 0$ *such that* $H_0 + V$ *has no bound states of energy* $E > E_0$.

Proof. By the Klein-Zemach theorem (Theorem I.23), $\lim_{\kappa \to \infty} \|R_\kappa\| = 0$. Thus, for some E_0, $\|R_\kappa\| < 1$ for $\kappa^2 > E_0$. Then, since bound states imply solutions of $R_\kappa \psi = \psi$, there are no positive energy bound states for $E > E_0$. ∎

Using a method related to one used by Grossman and Wu [45] to show there are only finitely many negative energy states when $V \epsilon R$ and short range, we prove:

THEOREM III.14. *Let* $V \epsilon R$ *and short range. Then* $H_0 + V$ *has at worst finitely many bound states of positive energy.*

Proof. By our discussion in Section I.5, the kernel R_κ has an analytic continuation into a neighborhood of the real axis. Thus, by the analytic Fredholm theorem, the places where $R_\kappa \psi = \psi$ has a solution (for κ real) form a discrete set. By Theorem III.13, this set can't extend to arbitrarily large κ; thus, the set of positive energy eigenvalues is finite. Since each R_κ is compact, the multiplicity of each is also finite. ∎

Remark

One might hope to prove the above theorem when $V \epsilon R$ by using the fact that R_κ is analytic in the upper half plane and continuous on the real line; however, there exist T_κ which obey this condition and which have solutions of $T_\kappa \psi = \psi$ when κ lies in a set with an accumulation point.[13] However, using the edge of the wedge theorem, we can show that the set of positive energy eigenvalues is nowhere dense.[14]

Note added in proof: For the case of V central and slightly better behaved than $V \epsilon R$, Ciafaloni and Menotti [146] have proven a radial integral equation analogous to (III.3) is equivalent to the radial Schrodinger equation for $E < 0$. They have used this (and related equations) to study anti-bound and resonance states.

---

13 There exist functions f analytic in the upper plane; continuous approaching the real line with zeros on the real line possessing an accumulation point. Let $T_\kappa = [f(\kappa) + 1]P$ where P is a finite-dimensional projection. To construct such an f, consider the function $g(z) = (\sin z)\, e^{-z^2}$ in the strip $0 \leq \text{Im } z \leq 1$. It is continuous on the closed strip and goes to 0 uniformly at ∞ in the strip. By a conformal mapping, we can map the strip onto the upper half plane with ∞ going into 0. Under this map g(z) goes into such an f since g has an accumulation point of zeros at ∞. I should like to thank J. Kohn for suggesting this construction.

14 We will actually see in Section IV. 5 that the set of points where $R_\kappa \phi = \phi$ has a solution is a closed set of measure 0.

CHAPTER IV

TIME-DEPENDENT SCATTERING THEORY

IV.1. *Introduction*

Scattering theory is a subject with a long and complicated history. It is only in the last decade and a half that its results have been put in a mathematically precise form. These mathematical developments have been accompanied by clarifications in the physical foundations of the theory.[1] The result is a conceptual picture quite appealing from the pedagogical point of view. To put this picture into perspective, let us first summarize its historical development.

The time-honored approach to scattering via wave functions that are asymptotically of the form $e^{i\vec{k}\cdot\vec{r}} + r^{-1} f(\theta) e^{ikr}$ is of a nature that "wouldn't convince an educated first grader."[2] In any attempt to develop a more satisfactory approach, one has to either tackle continuum (non-normalizable) states head on, or else deal completely with normalized states (wave packets). Surprisingly, the latter approach, which is more natural from a physical point of view, was not adopted (with the exception of Friedrichs [39]) until the works of Jauch [61] and Cook [24]. Previous authors who did work with wave packets (e.g., Hack [46], Moses [90])

---

[1] For a demonstration of the clarity of the more mathematical approach, compare Hack's 1958 paper [47] with his 1954 paper [46].

[2] In the words of M. L. Goldberger (in Physics 572, Spring term, 1968, Princeton University).

couldn't resist the temptation to expand in continuum states and secretly deal with them anyways.[3]

The modern era of scattering theory was ushered in by the classic works of Møller [89], Lippmann-Schwinger [86] and Gell-Mann and Goldberger [41]. Møller introduced the wave operators which we will discuss in detail shortly. These were treated as some sort of limit of $e^{+iHt}\, e^{-iH_0 t}$ (as $\to \pm\infty$). Because people insisted on applying these operators to continuum states, the notion of adiabatic switching of the scattering interaction was used – this idea of replacing V with $V\, e^{-\varepsilon|t|}$ is one of the most unphysical ideas in the history of physics. Gell-Mann and Goldberger appear to have been the first authors[4] to consider honestly in what sense $e^{+iHt}\, e^{-iH_0 t}$ has a limit as $t \to \pm\infty$. They proposed that the limits exist in the sense of Abel (in Section V.2, we will discuss Abelian limits and the Gell-Mann, Goldberger type of limit).

It was Cook [24] and independently Jauch [61] who first proposed that the limits $e^{+iHt}\, e^{-iH_0 t}$ should exist in the strong operator topology for H physical. Under the assumption that such a limit exists, Jauch derived various properties of the limit and of the S-matrix. Cook actually proved the limit exists when $H = H_0 + V$ with $V \epsilon L^2$. Jauch and Zinnes [62] showed $V \epsilon L^2$ was not necessary; in particular, they proved the limit existed for $V = r^{-\beta}$; $1 < \beta < \frac{3}{2}$. Hack [47] extended Cook's result to in-

---

[3] In all fairness to this approach, it must be admitted that continuum states are extremely useful for computational purposes and for proving general properties of the S-matrix; in fact, we use them extensively in Chapter VI. Nevertheless, I continue to insist that in discussing the physics behind scattering, and in particular, in defining the S-matrix, continuum states have no place.

[4] Again, with the exception of Friedrichs in his 1948 paper [39], which essentially proved the existence of the limits where $H = H_0 + V$ with certain unphysical V's. The crucial point is that he found the correct interpretation of the limit; namely strong operator convergence (on normalizable states).

clude $V = 0(r^{-1-\varepsilon})$ at ∞ (see Section IV.2 and also Kuroda [79], Brownell [8]). Since we know that the Coulomb force introduces infinite phase shifts, Hack's result appears to be the best possible with regard to behavior at ∞.[5]

After Cook's paper and simultaneous to the work of Hack and Jauch-Zinnes, Brenig and Haag [9] cast all of scattering theory into the more mathematical framework made possible by the existence of the limits of $e^{+iHt}e^{-iH_0t}$. This paper was crucial since its lucid presentation made the results known to a wide audience.

Having completed our bastardized history, let us return and discuss the physical and mathematical ideas behind the study of the limit

$$\lim_{t\to\pm\infty} e^{+iHt}e^{-iH_0t}.$$

We work in the Schrödinger picture. Thus the state whose time evolution is controlled by H and which is the vector Ψ at $t = 0$, is the vector $e^{-iHt}\Psi$ at time t. We will refer to $e^{-iHt}\Psi$ as "the interacting state Ψ at time t." We are thus dealing with an entire Schrödinger picture history.[6] By labeling the state with its $t = 0$ position, we are nearly in the Heisenberg picture, but it is useful to keep the distinction. Similarly, we refer to $e^{-iH_0t}\Phi$ as "the free state Φ at time t."

In order to describe scattering, we ask first what it means to say that the interacting state Ψ and the free state Φ look alike in the distant past. Clearly, a reasonable meaning for such a statement is

$$\lim_{t\to-\infty} \|e^{-iHt}\Psi - e^{-iH_0t}\Phi\| = 0.$$

---

5 However, by taking the phase shift into account, Dollard [26, 27] has succeeded in proving the existence of modified wave operators for the Coulomb and Coulomb-like problems.

6 Wightman [121] calls this sort of object a "state subspecie aeternitatis."

Since e^{-iHt} is unitary, this is equivalent[7] to $\lim_{t \to -\infty} \|\Psi - e^{+iHt} e^{-iH_o t}\Phi\| = 0$. Thus, if $\Psi = \lim_{t \to -\infty} e^{+iHt} e^{-iH_o t}\Phi$ exists, the interacting state Ψ looks like the free state Φ in the distant past. Since we expect there to be an interacting state which looks like any given free state in the past, we expect $\lim e^{+iHt} e^{-iH_o t}\Phi$ to exist for any Φ, i.e.

$$s - \lim_{t \to \mp \infty} e^{+iHt} e^{-iH_o t} = \Omega^{\pm} \qquad \text{(IV.1)}^8$$

should exist.

Before discussing scattering per se, let us make several remarks about the limit (IV.1):

(1) The norm limit will never exist. For an interacting state that involved a collision a million years ago will not reach the asymptotic past until t is very negative, more negative than is necessary for an interacting state colliding today.

(2) (first made by Hack [46]). We do not expect the norm limit $e^{+iH_o t} e^{-iHt}\Psi$ (compare with (IV-1)) to exist for all Ψ; for a bound (interacting) state does not look like any free state in the distant past.

(3) A weak operator limit in (IV-1) would not be suitable for $e^{+iH_o t} e^{-iHt} = (e^{+iHt} e^{-iH_o t})^*$ does have a weak limit [if (IV-1) exists] and thus $e^{+iH_o t} e^{-iHt}\Psi$ has a weak limit as $t \to -\infty$ when Ψ is a bound state. In fact it converges weakly to 0. Since such a bound state doesn't look like 0 in the past, weak limits are unsuitable.

---

7 This equivalence allows one to transfer considerations to an interaction picture. While such a procedure facilitates computations, it obscures the physics and was probably a cause of confusion in the early 1950's.

8 That the limits as $t \to \mp\infty$ are called $\Omega^{\pm}$ is a perversion coming from the $+i$ prescription. It is nearly universal among physicists (and almost non-existent among mathematicians). Since there are few nearly universal conventions in physics, it is our moral duty to preserve them all, even the perverted ones.

(4) Because strong limits are involved $\|\Omega^{\pm}\Phi\| = \|\Omega\|$, i.e., $\Omega^{\pm}$ is an isometry so $(\Omega^{\pm})^* \Omega^{\pm} = 1$.

(5) The wave operators $\Omega^{\pm}$ will not in general be onto, since bound states cannot be of the form $\Omega^{\pm}\Phi$. We thus call $\operatorname{ran} \Omega^{\pm} = \mathcal{H}_{\substack{in \\ out}} \cdot \mathcal{H}_{in}$ is the family of interacting states which asymptotically in the past look like free states.

(6) Thus $(\Omega^{\pm})(\Omega^{\pm})^* = P_{\substack{in \\ out}}$ where $P_{in(out)}$ is the projection on $\mathcal{H}_{in(out)}$.

Thus

$$(\Omega^{\pm})^* = \underset{t\to\mp\infty}{\text{w-lim}}\, e^{+iH_0t}\, e^{-iHt} \tag{IV.2}$$

preserves the size of vectors in $\mathcal{H}_{in(out)}$ and takes them into free states that they look like asymptotically in the past (future). States in $(\mathcal{H}_{\substack{in \\ out}})^{\perp}$ are annihilated by $(\Omega^{\pm}).^*$

(7) It is not hard to show that $e^{iHt}\Omega^{\pm} = \Omega^{\pm} e^{iH_0t}$ so $H\Omega^{\pm} = \Omega^{\pm} H_0$. In particular, H leaves the spaces $\mathcal{H}_{in(out)}$ invariant and $H \restriction \mathcal{H}_{in(out)}$ is unitarily equivalent to H_0.

If we think of the S-matrix as taking free states into free states, we are led naturally to the definition:[9]

$$S = (\Omega^-)^* (\Omega^+) \tag{IV.3}$$

for them $\langle\phi,S\psi\rangle = \langle\Omega^-\phi,\Omega^+\psi\rangle = \langle\phi^{out}, \psi^{in}\rangle$.

A close look at (IV.3) and Remark 6 leads one to realize that the unitary of the S-matrix is not merely a consequence of conservation of proba-

---

[9] In the multichannel case, things are complicated by "the non-orthogonality of channels"; as a result, a different definition is often used for multichannel scattering, cf. Jauch [61].

bility as the folklore leads us to believe. Clearly if $\mathcal{H}_{in}$ is not a subspace of $\mathcal{H}_{out}$, S will not preserve norms; physically the reason for this is clear – if $\mathcal{H}_{in} \not\subset \mathcal{H}_{out}$, there will be states that break up asymptotically in the past which do not in the future. Similarly $\mathcal{H}_{out} \subset \mathcal{H}_{in}$ is needed for S to be onto. Thus, unitarity of S requires at least $\mathcal{H}_{in} = \mathcal{H}_{out}$. That this is not a completely trivial question is shown by a contrived example of Kato and Kuroda [73]: this example presents a *non-local* potential in R^2 mildly *misbehaved at* ∞ for which $\mathcal{H}_{in} \neq \mathcal{H}_{out}$.

States that break up asymptotically in the past but not in the future, seem to defy physical interpretation, and so $\mathcal{H}_{in} = \mathcal{H}_{out}$ is an extremely attractive supposition. In fact, states which are in $(\mathcal{H}_{in})^{\perp}$ should be bound, because physically one can't see how an "unbound" state couldn't be asymptotically free. We are thus led naturally to isolate the following abstract properties:[10]

Weak Asymptotic Completeness (WAC)

$\mathcal{H}_{in} = \mathcal{H}_{out}$

Asymptotic Completeness (AC)

$\mathcal{H}_{in} = \mathcal{H}_{out} = (\mathcal{H}_{bound})^{\perp}$

Strong Asymptotic Completeness (SAC)

$\mathcal{H}_{in} = \mathcal{H}_{out} = (\mathcal{H}_{bound})^{\perp}$ and all bound states have non-positive energy.

We have isolated SAC because it seems to be difficult to prove AC without proving (or assuming!) facts about the positive energy spectrum along the way. In addition, SAC has the very simple form:

---

[10] The names weak and strong asymptotic completeness are not standard – they are an invention of this author.

$$\text{Ran } P_{(0,\infty)} = \mathcal{H}_{in} = \mathcal{H}_{out}$$

in terms of the spectral projection $P_{(0,\infty)}$ for H.

We will discuss the known results on (WAC), (AC) and (SAC) in later sections of this chapter. However, at this juncture, let us mention that Kuroda [79] provided the earliest proof of (WAC) for a large class of potentials and Ikebe [56] first proved (SAC) for a different (but large) class.

The various forms of asymptotic completeness can be put into sharper focus by looking at the classification of the spectral measures of a Hamiltonian. In the appendix (Theorem A.12), we see that given a self-adjoint Hamiltonian H on a Hilbert space $\mathcal{H}$, we can decompose

$$\mathcal{H} = \mathcal{H}_{a.c.} \oplus \mathcal{H}_{p.p.} \oplus \mathcal{H}_{sing}$$

where each of the spaces is left invariant by H and where $H \restriction \mathcal{H}_{a.c.}$ has absolutely continuous spectrum, $H \restriction \mathcal{H}_{p.p.}$ has a complete set of eigenvectors and $H \restriction \mathcal{H}_{sing}$ has continuous singular spectrum. Since $H \restriction \mathcal{H}_{in(out)}$ is unitarily equivalent to H_0 and H_0 has only absolutely continuous spectrum, we must have $\mathcal{H}_{in(out)} \subset \mathcal{H}_{a.c.}$. A natural way of realizing (WAC) is to have $\mathcal{H}_{in(out)}$ both equal to $\mathcal{H}_{a.c.}$. This situation was studied first in an abstract setting by Kato [67, 68] and in a physical setting by Kuroda [79] (this is how Kuroda proved (WAC) for $V \in L^1 \cap L^2$). We thus introduce a fourth and auxiliary form of asymptotic completeness:

Kato (Kuroda) – Completeness (KC)

$\mathcal{H}_{in} = \mathcal{H}_{out} = \mathcal{H}_{a.c.}$

It is clear that:

THEOREM IV.1. (SAC) => (AC) => (KC) => (WAC). ∎

One can appreciate the mathematical content of (AC) by seeing how it is stronger than (KC). By definition $\mathcal{H}_{bound} = \mathcal{H}_{p.p.}$ so (AC) implies

$\mathcal{H}_{sing} = 0$. Conversely (KC) implies $\mathcal{H}_{in}{}^{\perp} = \mathcal{H}_{a.c.}{}^{\perp} = (\mathcal{H}_{sing} \oplus \mathcal{H}_{bound})$ so (KC) $\wedge$ $(\mathcal{H}_{sing} = 0)$ implies (AC). We have thus shown:

THEOREM IV.2. (KC) $\wedge$ $(\mathcal{H}_{sing}\ 0)$ <=> (AC). ∎

Thus, while asymptotic completeness seems physically clear, it implies that H has no singular spectrum; since this is a deep mathematical result, (AC) is much more difficult to prove than (WAC) [or (KC)]. In fact, Aronszajn [2] has an example of a one-dimensional potential *very misbehaved* at ∞ for which $H = p^2 + V$ has singular spectrum; thus, the full form of (AC) is very deep indeed.

Having completed our review of the status of mathematical two-body scattering theory, let us discuss in what ways these results carry over to our class of potentials. Intuitively, it is the behavior at ∞ which is crucial to scattering and thus one would expect the change from L^2 locally to R locally to be inessential. This is more or less true. In Sections IV.2 and IV.3, we will show that the Cook-Hack and Kuroda results go over without change, i.e., if $V \epsilon R + L^{\infty}$ and $V \sim 0(r^{-1-\varepsilon})$ at ∞, then $\Omega^{\pm}$ exist and if $V \epsilon R \cap L^1$, the theory obeys (KC), and thus (WAC). In fact, by using more recent results of Kato, we can prove (KC) for any $V \epsilon R$ and $V \epsilon R + L^{\infty}$ bounded by $(|x|)^{-1-\varepsilon}$ at ∞. We study the approximation properties of $\Omega^{\pm}$ and the S-matrix in Section IV.4 and obtain a mild improvement even in the $V \epsilon L^1 \cap L^2$ case. In IV.5 and IV.6, we show that Ikebe's method carries over after modification on a set of measure 0. In particular, we present a new proof that (SAC) holds for $V \epsilon L^1 \cap R$ with $\|V\|_R < (4\pi)$. This is a special case[11] of a result of Kato [70].

---

11 Kato did not require $V \epsilon L^1$!

IV.2. *The Existence of the Møller Wave Operators*

In this section, we will show that if $V \epsilon R + L^\infty$ and $V = 0(r^{-1-\varepsilon})$, then the wave operators exist. The proof will involve a minor perturbation on the Cook-Hack proof [24, 47] of the analogous result for $V \epsilon L^2 + L^\infty$. These results are not new; they may be found in the paper of Kupsch and Sandas [78].[12] Cook's theorem on the existence when $V \epsilon L^2$ follows from the following sequence of lemmas whose proof is fairly simple:

LEMMA IV.3. *If* $\lim_{t \to -\infty} e^{+iHt} e^{-iH_0 t} \Phi$ *exists for* $\Phi \epsilon D$, *some dense set of* $\mathcal{H}$, *then it exists for all* $\Phi \epsilon \mathcal{H}$. ∎

LEMMA IV.4. *If* $\Phi(t)$ *is a function of* t *so that* $\frac{d}{dt}\Phi(t) \equiv \Phi'(t)$ *exists (in the sense of norm limits) and so that* $\int_{T_0}^\infty \|\Phi'(t)\|\, dt < \infty$, *then* $\lim_{t \to \infty} \Phi(t)$ *exists.* ∎

LEMMA IV.5. *Let* $\Psi \epsilon L^1 \cap L^2$. *Then for* $t \neq 0$:

$$(e^{-iH_0 t}\Psi)(x) = (4\pi i t)^{-3/2} \int d^3y \; e^{+i|x-y|^2/4t}\Phi(y). \blacksquare$$

LEMMA IV.6. *Let* $\Psi \epsilon L^1 \cap L^2$. *Then, for* $t \neq 0$, $e^{-iH_0 t}\Psi \epsilon L^\infty$ *and* $\|e^{-iH_0 t}\Psi\|_\infty \leq C|t|^{-3/2}$ *for a* Ψ*-dependent constant* C.

Proof. Follows from Lemma IV.5. ∎

To see how these lemmas conspire to yield Cook's result, we prove (following Dolard [27]):

THEOREM IV.7. (Cook). *If* $V \epsilon L^2$, $\Omega^\pm$ *exist.*

---

[12] They, however, do not reach any conclusion about the definition of $H_0 + V$ when $V \epsilon L^2 + L^\infty$, even when $V \epsilon R + L^\infty$. They merely suppose some definition of H is given which is the sum of H_0 and V for functions with support in a region where $V \epsilon (L^2)_{loc}$.

Proof. Let $\Phi \in \mathcal{S} \subset D(H_o) = D(H)$. Let $\Phi(t) = e^{+iHt} e^{-iH_o t} \Phi$. Then $\Phi'(t) = ie^{+iHt}(H - H_o) e^{-iH_o t} \Phi$ so

$$\|\Phi'(t)\| \leq \|Ve^{-iH_o t}\Phi\| \leq \|V\|_2 \|e^{-iH_o t}\Phi\|_\infty \leq C\|V\|_2 |t|^{-3/2}$$

(by Lemma IV.6). Thus $\int_{\pm 1}^{\pm\infty} \|\Phi'(t)\| \, dt < \infty$ so $\lim_{t \to \pm\infty} e^{+iHt} e^{-iH_o t} \Phi$ exists when $\Phi \in \mathcal{S}$ by Lemma IV.4. Since $\mathcal{S}$ is dense in L^2, $\Omega^\pm$ exist by Lemma IV.3. ∎

Cook's result can be extended to Hack's:

THEOREM IV.8 (Hack).[13] *Let* $(1 + |x|)^{-\frac{1}{2}+\varepsilon} V \in L^2$ [*for example, suppose* $V \underset{x \to \infty}{=} 0(|x|^{-1-\delta})$]. *Then* $\Omega^\pm$ *exist.*

Proof. As above one shows $\|Ve^{-iH_o t}\Phi\| \in L^1(t)$. One can take Φ to be Gaussian, do an explicit computation [see Kuroda [79] or Kato [T17], pp. 534-535] and use the fact that finite sums of Gaussians are dense. Alternately, following Kupsch and Sandas [78], one can improve Lemma IV.6 for $\Psi \in \mathcal{S}$ and any α to read

$$|(e^{-iH_o t}\Psi)(x)| \leq C_{\alpha,\Psi}(t/x)^\alpha \, t^{-3/2}$$

for $t > 1$ for some C dependent on α and Ψ only and use that. ∎

In proving an analogous result for $V \in R + L^\infty$, we must take more care; for $e^{+iHt} e^{-iH_o t} \Psi$ may not have a norm-limit derivative since D(H) and $D(H_o)$ may have little in common. However, a simple trick avoids problems:

---

[13] As stated, this theorem is due to Kuroda [79] and is a slight improvement of Hack's theorem which required $V = 0(|x|^{-1-\delta})$ at ∞.

THEOREM IV.9. *Let* $V \epsilon R + L^{\infty}$ *and suppose that*

$$\int_{|r|>x} |V(r)|^2 (1+|r|)^{-1+\varepsilon} d^3r < \infty$$

for some $\varepsilon > 0$ *and some* x. *Then* $\Omega^{\pm}$ *exist.*

Remarks

1. This result includes the Cook-Hack theorem as a special case. It cannot be improved much at ∞ without running into the Coulomb problem for which no unmodified wave operators exist nor can it be improved much at finite singularities without running into Nelson's phenomena for r^{-2} potentials, [91].[14]

2. This result allows one to extend the Jauch-Zinnes examples to $V = r^{-\beta}$ $(1 < \beta < 2)$. Such an extension is actually obtained by Green and Lanford [43] where *central* potentials with $V \sim r^{-2+\varepsilon}$ near $r = 0$ are allowed.

3. This theorem is the main result of the paper of Kupsch and Sandas [78].

Proof (Kupsch-Sandas [78]). Pick a C^{∞} positive function $F(r) \leq 1$ which is $\equiv 1$ on $\{\vec{r} \mid |\vec{r}| < x\}$ and which vanishes for $|\vec{r}| > 2x$. Let $\Phi \epsilon \mathcal{S}$. First decompose:

$$e^{+iHt} e^{-iH_0t}\Phi = e^{+iHt}F e^{-iH_0t}\Phi + e^{+iHt}(1-F) e^{-iH_0t}\phi.$$

The first term goes to 0 as $t \to \infty$, for:

$$\|Fe^{-iH_0t}\Phi\|^2 \leq [\text{volume } \{\vec{r} \mid |\vec{r}| < 2x\}] \; \|e^{-iH_0t}\Phi\|_{\infty}^2 \leq C|t|^{-3} \to 0 \text{ as } t \to \pm\infty.$$

---

14 However positive singularities worse than r^{-2} can occur. In that case a proof essentially identical to the one below will work and prove existence of $\Omega^{\pm}$.

Thus, to show the wave operators exist, we must only show

$$e^{+iHt}(1-F)\, e^{-iH_o t}\Phi \equiv \Phi(t)$$

has a norm limit as $|t| \to \infty$ and we use Lemma IV.4 to prove this. Since the integral in the hypothesis of the theorem is finite, we see that any function in $\mathcal{S}$ vanishing outside of $\{\vec{r} \mid |\vec{r}| < x\}$ is in $D(H_o) \cap D(V) \subset D(H)$. Thus $\Phi'(t)$ exists, in fact:

$$\Phi'(t) = ie^{+iHt}\{H(1-F) - (1-F)H_o\}\, e^{-iH_o t}\Phi$$

$$= ie^{+iHt}[H_o,(1-F)]\, e^{-iH_o t}\Phi + ie^{+iHt}V(1-F)\, e^{-iH_o t}\Phi.$$

The second term has norm in $L^1(t)$ by the argument in the proof of Theorem IV.8. To handle the first term, we compute:

$$[H_o, 1-F] = \Delta F + 2i\vec{\nabla} F \cdot \vec{p}.$$

Thus the first term has a norm bounded by

$$\|(-\Delta F)\, e^{-iH_o t}\Phi\| + 2\|\ \vec{F} \cdot e^{-iH_o t}(\vec{p}\Phi)\|$$

each of which goes as $|t|^{-3/2}$ $(t \to \pm\infty)$. Thus $\int_{\pm 1}^{\pm\infty} \|\Phi'(t)\|\, dt < \infty$ so the wave operators exist. ∎

IV.3. *Unitarity of the S-Matrix*

In this section, we will show that Kuroda's proof of (WAC) [79] for $V \epsilon L^2 \cap L^1$ carries over with minor changes to cover $V \epsilon R \cap L^1$. We will then show how one can prove (KC) [and thus (WAC)] for any V in R and a large class in $R + L^\infty$. The crucial mathematical result which has been abstracted from Kuroda's approach is:

LEMMA IV.10. *Let* H *be an operator such that* $(H-E)^{-1} - (H_o-E)^{-1}$ *is trace class for some* E. *Then the wave operators exist and are* (KC).

Proof. See Kato [T17], pp. 535-546, esp. Theorem X.4.8. ∎

This lemma is due to Birman [7]. The general approach relied on ideas of Putnam [96], Kato [67, 68] and Rosenblum [99]. Kuroda's theorem is an immediate consequence of this lemma (see Kato [T17], pp. 546-547 or below). One also has:

THEOREM IV.11. *Let* $V \in R \cap L^1$. *Then* $\Omega^{\pm}$ *exist and obey* (KC). *In particular, the S-matrix is for such a potential is unitary.*

Proof. Let $E < 0$, $E \notin \text{spec}(H)$. By Theorem II.37, $(H-E)^{-1}-(H_0-E)^{-1}$ is trace class. Thus Lemma IV.10 implies the result. ∎

Remarks

1. This result (which can also be proven using powerful new techniques of Kato and Kuroda) includes Kuroda's $L^1 \cap L^2$ result as a special case.

2. In the L^2-case, Kuroda's $L^1 \cap L^2$ potentials are a subclass of Cook's L^2 potentials; however, $R \cap L^1$ is not quite a subclass of the potentials treated by Kupsch and Sandas.

3. Footnote 3 on page 546 of Kato's book [T17] actually contains the one-dimensional analogue of Theorem IV.11. In one dimension, the analogue of $V \in R$ is $V \in L^1$ as we have seen, so that the analogue of $L^1 \cap R$ is merely L^1.

The extension of Theorem IV.11 to a larger class of V's depends on extending the notion of wave operators.

DEFINITION. Let H_1, H_2 be two self-adjoint operators with absolutely continuous spaces $\mathcal{H}^{(1)}_{a.c.}$, $\mathcal{H}^{(2)}_{a.c.}$. We say the wave operators $\Omega^{\pm}(H_2,H_1)$ exist if and only if $\lim_{t\to\mp\infty} e^{+iH_2t}e^{-iH_1t}\psi$ exist for each $\psi \in \mathcal{H}^{(1)}_{a.c.}$. We then define $\Omega^{\pm}(H_2,H_1)$ on $\mathcal{H}^{(1)}_{a.c.}$ by this limit and on $(\mathcal{H}^{(1)}_{a.c.})^{\perp}$ as 0. If the wave operators $\Omega^{\pm}(H_2,H_1)$ have range $\mathcal{H}^{(2)}_{a.c.}$, we say they are compl

It is rather straight-forward to prove:

LEMMA IV.12. *Let* $\Omega^{\pm}(H_2,H_1)$ *exist. Then, they are complete if and only if* $\Omega^{\pm}(H_1,H_2)$ *exist.*

(b) *If* $\Omega^{\pm}(H_2,H_1)$ *and* $\Omega^{\pm}(H_3,H_2)$ *exist, then so does* $\Omega^{\pm}(H_3,H_1)$ *and* $\Omega^{\pm}(H_3,H_1) = \Omega^{\pm}(H_3,H_2)\,\Omega^{\pm}(H_2,H_1)$.

Moreover, if $\Omega^{\pm}(H_2,H_1)$ *and* $\Omega^{\pm}(H_3,H_2)$ *are both complete, then so is* $\Omega^{\pm}(H_3,H_1)$.

Proof. See Kato [T17], p. 532. ∎

Moreover, the analogue of Lemma IV.10 holds:

LEMMA IV.13 (Kato-Birman theorem). *If* $(H_2-E)^{-1} - (H_1-E)^{-1}$ *is trace class, then* $\Omega^{\pm}(H_2,H_1)$ *exist and are* (KC). ∎

We also abstract the following two results from the literature:

THEOREM IV.14 (Kato [70]). *Let* $V \epsilon R$ *with* $\|V\|_R < 4\pi$. *Then* $\Omega^{\pm}(H_0 + V, H_0)$ *exist and are* (KC). [In fact they are unitary.] ∎

THEOREM IV.15 (Kato [72]). *Let* $|V(x)| < C(1+|x|)^{-1-\varepsilon}$. *Then* $\Omega^{\pm}(H_0 + V, H_0)$ *exist and are* (KC). ∎

What we will prove is:

THEOREM IV.16. *Let* $V_1, V_2 \epsilon R + L^{\infty}$ *and suppose that:*

(a) $\Omega^{\pm}(H_0 + V_1, H_0)$ *exists and obeys* (KC).

(b) $V_1 - V_2 \epsilon L^1 \cap R$.

Then $\Omega^{\pm}(H_0 + V_2, H_0)$ *exists and obeys* (KC).

Proof. By Lemma IV.12, it is sufficient to show $\Omega^{\pm}(H_0 + V_2, H_0 + V_1)$ exists and obeys (KC). By Lemma IV.13, this will follow if we can show

$(H_o + V_2 - E)^{-1} - (H_o + V_2 - E)^{-1}$ is trace class for some E. Letting $x_i = (H_o - E)^{-1/2} V_i (H_o - E)^{-1/2}$ $(i = 1,2)$ and picking $E \notin \sigma(H_1) \cap \sigma(H_2)$, we see that

$$(H_2 - E)^{-1} - (H_1 - E)^{-1} = (H_o - E)^{-1/2}[(1+x_2)^{-1} - (1+x_1)^{-1}](H_o - E)^{-1/2}$$

$$= (H_o - E)^{-1/2}(1+x_2)^{-1}(x_1 - x_2)(1+x_1)^{-1}(H_o - E)^{-1/2}$$

$$= (H_o - E)^{-1/2}\{1-(1+x_2)^{-1}x_2\}(x_1 - x_2)\{1 - x_1(1+x_1)^{-1}\}$$

$$(H_o - E) - 1/2$$

$$= (I) + (II) + (III) + (IV).$$

Now

$$(I) = (H_o - E)^{-1/2}(x_1 - x_2)(H_o - E)^{-1/2} = (H_o - E)^{-1}(V_1 - V_2)(H_o - E)^{-1}$$

is trace class since $V_1 - V_2 \in L^1$.

To treat

$(II) = (H_o - E)^{-1/2}(1+x_2)^{-1}x_2\,(x_1 - x_2)x_1\,(1+x_1)^{-1}(H_o - E)^{-1/2}$ we write $V_i = B_i + R_i$ where $B_i \in L^\infty$, $R_i \in R$. It is certainly sufficient to show $x_2(x_2 - x_1)x_1$ is trace class and so we write:

$$x_2(x_2 - x_1)x_1 = (a) + (b) + (c) + (d)$$

where

$(a) = (E - H_o)^{-1/2} B_2[(E - H_o)^{-1}(V_2 - V_1)(E - H_o)^{-1}] B_1 (E - H_o)^{-1/2}$ is trace class since B_1, B_2 are bounded and [...] is trace class as above.

$(b) = [(E - H_o)^{-1/2} R_2 (E - H_o)^{-1/2}](E - H_o)^{-1/2}(V_1 - V_2)(E - H_o)^{-1/2}$ $[(E - H_o)^{-1/2} R_1 (E - H_o)^{-1/2}]$ is trace class since the [...] terms are both Hilbert-Schmidt. (c) and (d) are handled as (b) was and thus (II) is trace-class. The cross terms (III) and (IV) are handled by means identical to those used to treat (II). ∎

Combining Theorems IV.16 and IV.14, we obtain

THEOREM IV.17. *Let* $V \in R$. *Then* $\Omega^{\pm}$ [*the physical wave operators* $\Omega^{\pm}(H_0 + V, H_0)$] *exist and obey* (KC).

Proof. Since $\int d^3x\, d^3y\, |V(x)|\, |x-y|^{-2}\, |V(y)| < \infty$, we can pick R_0 so that $\int_{|x|,|y|>R_0} d^3x\, d^3y\, |V(x)|\, |V(y)|\, |x-y|^{-2} < (4\pi)^2$. Let $W(x)$ be defined by

$$W(x) = \begin{cases} V(x) & |x| \geq R_0 \\ 0 & |x| < R_0 \end{cases} \tag{IV.4}$$

Then $W \in R$ and $\|W\|_R < (4\pi)$ so $\Omega^{\pm}(H_0 + W, H_0)$ exist and are (KC), by Theorem IV.14 (Kato). But $V\text{-}W$ is in R and has compact support and thus it is in $R \cap L^1$. As a result, the physical $\Omega^{\pm}$ exist and are (KC) by Theo-IV.16. ∎

And combining Theorems IV.16 and IV.15 yields:

THEOREM IV.18. *Let* $V \in R + L^{\infty}$ *and suppose for some* $\varepsilon > 0$, R_0 *and* C, $|x| > R_0$ *implies* $|V(x)| < C|x|^{-1-\varepsilon}$. *Then* $\Omega^{\pm}$ *exist and are* (KC).

Proof. Let $W(x)$ be defined by (IV.4). Then $|W(x)| < C(1 + |x|)^{-1-\varepsilon}$ so $\Omega^{\pm}(H_0 + W, H_0)$ exist by Theorem IV.15. $W\text{-}V \in L^{\infty} + R$ and of compact support is thus in $L^1 \cap R$. Thus $\Omega^{\pm}$ exist and are (KC) by Theorem IV.16. ∎

Remarks

1. In terms of *behavior at* ∞, IV.17 is more or less equivalent to Ikebe's class [56]. [But, of course, Ikebe proved (SAC), not merely (KC).] However Ikebe needs smoothness and $V \in L^2$.

2. According to our discussion following Theorem IV.9 (Remark 1), Theorem IV.18 is the best possible expected result for (KC).

3. Theorem IV.17 is not quite implied by Theorem IV.18 since IV.17 does not even require $V \to 0$ at ∞.

4. The moral of Theorem IV.16 is that behavior at ∞ is the crucial aspect of existence and unitarity of S (not an unexpected fact!). Once one can handle some behavior at infinity when V is "nice" in finite regions, one can handle this behavior at ∞ for any $V \epsilon R + L^{\infty}$. In particular since Kato has done the hard work [70, 72], we have leaned back, turned the crank of Theorem IV.16 and extended his results to maximally nice ones (with respect to KC – not AC or SAC).

Appendix to IV.3. Modified (KC) for Rollnik potentials with Coulombic tails.

As another support for the philosophy we expounded at the end of the last section, we consider scattering in a potential $V = cr^{-1} + \tilde{V}$ with $\tilde{V} \epsilon R \cap L^1$. As we have already remarked, infinite phase shifts suggest that the usual wave operators will not exist in this case or in the purely Coulombic case. Dollard has proven the following substitute however [26, 27]:

LEMMA IV.19. *Let* $H_c = -\Delta + cr^{-1}$. *Let*

$$H_{co}(t) = -\Delta t + \tfrac{1}{2}\varepsilon(t)\, c(-\Delta)^{-\frac{1}{2}} \ln[-4|t|\Delta] \tag{IV.5}$$

where $\varepsilon(t) = t/|t|$. *Then* $\Omega_D^{\pm} \equiv \text{S-}\lim_{t \to \mp\infty} e^{+iH_c t} e^{-iH_{co}(t)}$ *exist and* $\operatorname{Ran} \Omega_D^+ = \operatorname{Ran} \Omega_D^- = \mathcal{H}_{a.c.}(H_c)$.

Proof. See [26, 27]. For motivation on the choice of $H_{co}(t)$, see [26], pp. 114-122. ∎

$H_{co}(t)$ is a suitable substitute for H_0 for three reasons:

(a) If one uses the "standard" Coulombic continuum wave functions, $\psi_{\vec{k}}^{+}$ then:

$$\Omega_D^+ \left(\int d^2k\; f(k)\; e^{+ik\cdot x} \right) = \int d^3k\; f(k)\; \psi_k^{(+)}(x).$$

(b) $e^{iH_c t}\,\Omega_D^{\pm} = \Omega_D^{\pm}\,e^{iH_o t}$.

(c) While $e^{-iH_{co}(t)}\psi$ does not define a state with free dynamics, as $t \to \pm\infty$, $e^{-iH_{co}(t)}\psi$ has a probability distribution which approaches that of $e^{-iHt}\psi$ (i.e., $\lim_{t\to\pm\infty} \int \| (e^{-iH_{co}(t)}f)(x)|^2 - |(e^{-iH_o t}f)(x)|^2 |\, d^3x = 0$).

Thus, $\Omega^-\phi$ describes an interacting state which in the past has an asymptotic *probability distribution* identical to that of the free state ϕ in x space. Since $e^{-iH_{co}(t)}$ is diagonal in p-space, the p-space *distributions* are also identical. We can now use Theorem IV.16 to prove:

THEOREM IV.20. *Let* $H = -\Delta + V$ *where* $V = cr^{-1} + \tilde{V}$; $\tilde{V} \epsilon R \cap L^1$. *Let* $H_{co}(t)$ *be given by (IV.5). Then*

$$\text{S-}\lim_{t\to\pm\infty} e^{+iHt}\, e^{-iH_{co}(t)} = \Omega_D^{\mp}$$

exists and $\text{Ran}\,\Omega_D^{\pm} = \mathcal{H}_{a.c.}(H)$. *In particular, the modified S-matrix:*

$$S_D = (\Omega_D^-)^*\,\Omega_D^+$$

is unitary.

Remark

Dollard also showed that $\Omega_D^{\pm}$ exist when $\tilde{V} \epsilon L^2$ by a modification of Cook's argument.

Proof. By Lemma IV.19, it is sufficient to prove $\Omega^{\pm}(H, H_o + cr^{-1})$ exist and are complete. This follows from the proof of Theorem IV.16. ∎

IV.4. *Approximation Theorems*

Kuroda has obtained the following result as a corollary of his treatment of existence and (KC) when $V \epsilon L^1 \cap L^2$ [79]:

THEOREM IV.21. (Kuroda). *Let* V_n, $V \epsilon L^1 \cap L^2$ *and suppose* $V_n \to V$ *in* L^1 *and that* $\{\|V_n\|_2\}$ *is bounded. Let* $\Omega_n^\pm$, $\Omega^\pm$ *be the wave operators for* $H_o + V_n$, $H_o + V$ *and let* S_n, S *be the S-matrices. Then:*

$$\Omega_n^\pm \to \Omega^\pm \quad \textit{(strongly)}$$

$$S_n \to S \quad \textit{(strongly)}$$

Proof. See Kuroda [79]; alternately, this will be a corollary of Theorem IV.23 below. ∎

Kuroda's method has been abstracted by Kato to the following theorem:

LEMMA IV.22. *Let* H_n, H, H_o *be self-adjoint operators where* H_o *has only absolutely continuous spectrum. Suppose for fixed* E,

$$A_n = (H_n - E)^{-1} - (H_o - E)^{-1}$$

and $A = (H-E)^{-1} - (H_o - E)^{-1}$ *are all trace class and suppose* $\|A_n - A\|_1 \to 0$ *(where* $\|A\|_1 = \operatorname{tr} \sqrt{(A^*A)}$. *Then* $\Omega_n^\pm \to \Omega^\pm$ *(strongly).*

Proof. See Kato [T17], pp. 551; esp. remark 4.17. ∎

Using Lemma IV.22, we will extend Kuroda's result to $L^1 \cap R$:

THEOREM IV.23. *Let* V_n, $V \epsilon R \cap L^1$ *and suppose* $V_n \to V$ *in* L^1 *and in* R. *Then* $\Omega_n^\pm \to \Omega^\pm$ *(strongly) and* $S_n \to S$ *(strongly).*

Proof. We know $\|V_n - V\|_R \to 0$ so we can pick E so negative that $\|(H_o-E)^{-1/2} V_n (H_o-E)^{-1/2}\| < 1/2$ for all n sufficiently large. For such an n, all the $(H_n-E)^{-1} - (H_o-E)^{-1}$ are trace class since V, $V_n \epsilon L^1 \cap R$. We first show $\|(H_n-E)^{-1} - (H-E)^{-1}\|_1 \to 0$. For let

$$x = (H_o-E)^{-1/2} V (H_o-E)^{-1/2}; \quad x_n = (H_o-E)^{-1/2} V_n (H_o-E)^{-1/2}.$$

Using the Tiktopoulos formula, we have:

$$(H_n-E)^{-1} - (H-E)^{-1} = (H_0-E)^{-1/2} [(1+x_n)^{-1} - (1+x)^{-1}] (H_0-E)^{-1/2}$$

$$= (H_0-E)^{-1/2} (1+x_n)^{-1} (x-x_n) (1+x)^{-1} (H_0-E)^{-1/2}$$

$$= (I) + (II) + (III) + (IV),$$

where (I) – (IV) are the terms obtained by using $(1 + x)^{-1} = 1 - x(1 + x)^{-1}$ and the analogous formula for x_n. Then:

$$(I) = (H_0-E)^{-1}(V-V_n) (H_0-E)^{-1}$$

has trace norm bounded by:

$$\int dx\, dy\, |V(x) - V_n(x)| \frac{e^{-2k|x-y|}}{|x-y|^2} \leq C\|V-V_n\|_1$$

which goes to 0 since $V_n \to V$ in L^1.

$$(II) = (H_0-E)^{-1/2} x_n(1+x_n)^{-1} (x-x_n) x(1+x)^{-1} (H_0-E)^{-1/2}$$

has a trace norm bounded by:

$$\|(H_0-E)^{-1/2}\|^2 \|(1+x_n)^{-1}\| \|x_n\| \|(1+x)^{-1}\| \|x-x_n\|_{H.S.} \|x\|_{H.S.}$$

where $\|A\|^2_{H.S.} = \operatorname{Tr}(A^*A)$.

Since $V \epsilon R$, $\|x\|_{H.S.} < \infty$ and since $V_n \to V$ in R, $\|x-x_n\|_{H.S.} \to 0$, so (II) has trace norm going to zero.

Terms (III) and (IV) are treated analogously to (II).

Thus $\|(H_n-E)^{-1} - (H-E)^{-1}\|_1 \to 0$ so $\Omega_n^{\pm} \to \Omega^{\pm}$ (strongly) by Lemma IV.22. Thus

$$(\phi,S_n\psi) = (\Omega_n^-\phi,\Omega_n^+\psi) \to (\Omega^-\phi,\Omega^+\psi) = (\phi,S\psi)$$

so that $S_n \to S$ (weakly). Since all the S operators are unitary (by Theorem IV.11), $S_n \to S$ strongly. ∎

Remarks

1. One might expect $V_n \to V$ in L^1 and $\|V_n\|_R$ bounded to be sufficient for convergence. However, when we obtain a deeper understanding of the convergence theorem in the next section it will be clear that $\|\ \|_R$ *convergence* seems needed. For this will imply $(V_n)^{1/2} (E-H_0)^{-1} V_n^{1/2}$ converges in H.S. norm to $V^{1/2} (E-H_0)^{-1} V^{1/2}$ as the following shows: Given any subsequence V_{n_i}, we can as in Section I.2 find a sub-subsequence with $V_{n_{i(j)}}(x) \to V(x)$ pointwise and with $\frac{|V_{n_{i(j)}}(x)|\,|V_{n_{i(j)}}(y)|}{|x-y|^2} \le \frac{|W(x)|\,|W(y)|}{|x-y|^2}$ for some $W \in R$. Then, by the dominated convergence theorem

$$\int \left[\frac{V_{n_{i(j)}}^{1/2}(x)\, V_{n_{i(j)}}^{1/2}(y)}{|x-y|} - \frac{V^{1/2}(x)\, V^{1/2}(y)}{|x-y|} \right]^2 d^3x\, d^3y \to 0.$$

Thus, any subsequence, has a sub-subsequence converging in H.-S. norm to the V-kernel, i.e., the original sequence converges.

2. Since we require $V_n \to V$ (L^1 *and* R) and Kuroda only needs $V_n \to V(L^1)$ and $\|V_n\|_2$ bounded, a sloppy look would seem to indicate that Theorem IV.23 is not an extension of Kuroda's theorem. However, by Theorem I.4, $\|V-V_n\|_R \le C\|V-V_n\|_{L^1}^{1/3} \|V-V_n\|_{L^2}^{2/3}$ so Kuroda's conditions do imply $V_n \to V$ ($\|\ \|_R$). As a result, Theorem IV.23 *does* imply IV.21.

3. Even in the case that all V_n, $V \in L^1 \cap L^2$, Theorem IV.23 is (slightly) stronger than Theorem IV.21. For example, let

$$V_n^{(\alpha)}(x) = \begin{cases} n^\alpha; \ |x| < n^{-1} \\ 0 \text{ otherwise.} \end{cases}$$

Then, Theorem IV.21 says $S_n^{(\alpha)} \to 1$ for α fixed with $\alpha \le \frac{3}{2}$ while Theorem IV.23 implies this result for $\alpha < 2$.

IV.5. *Eigenfunction Expansions*

In Ikebe's proof of asymptotic completeness for Hölder continuous potentials, which are locally L^2 and $0(|x|^{-2-\varepsilon})$ at infinity [56], eigenfunction expansions played an essential role. In this section, we attempt to carry Ikebe's method over to an arbitrary $L^1 \cap R$ potential; by brushing a set of measure zero under the rug, we succeed – up to a point. Our approach will make essential use of the compactness of various operators and thus of the three dimensional nature of the problem. Bertero et. al., extend Ikebe's results to a class of "non-local" potentials. Thöe [108] has extended Ikebe's results to R^n; it would be interesting to see to what extent one could carry eigenfunction expansions over to R^n for potentials defining Hamiltonians by a forms method.[15] We make use of a technical result (Lemma IV.24) whose proof we defer to an appendix.

DEFINITION. Given $V \epsilon L^1 \cap R$, we say $|\kappa|^2$ is an exceptional value if and only if the kernel

$$R_{|\kappa|}(x,y) = -\frac{1}{4\pi} V_{||}^{1/2}(x) \frac{e^{i|\kappa|\,|x-y|}}{|x-y|} V^{1/2}(y) \tag{I.17}$$

has a homogeneous L^2 solution $(\psi = R_{|\kappa|}\psi)$. We denote the family of exceptional points by $\mathcal{E}$ and its complement (in $(0,\infty)$) by $\mathcal{R}$.

LEMMA IV.24. *$\mathcal{E}$ is a closed set of Lebesgue measure zero.* ∎

---

[15] The techniques of Kuroda which we discuss in a bibliographic note probably yield results for this case.

We actually expect this set to be empty, but we are unable to prove this in general. We remark that in two cases, we know something about $\mathcal{E}$. If $\|V\|_R < 4\pi$, $\mathcal{E} = \phi$ and if $V = 0(e^{-a|x|})$ at ∞, then $\mathcal{E}$ is discrete.

The basic theorem of this section will be proven in a series of lemmas:

THEOREM IV.25. *Let* $V \epsilon L^1 \cap R$. *Let* $E_{a.c.}$ *and* E_{sing} *be the projections on* $\mathcal{H}_{a.c.}$ *and* $\mathcal{H}_{sing}$ *for* $H = H_0 + V$. *Then, there exists a family of functions* $\phi(x,k)$ *defined for all* $k \epsilon R^3$ *with* $|k|^2 \epsilon \mathcal{E}$ *so that:*

(i) *For any* $f \epsilon L^2$ $\hat{f}(k) = \text{l.i.m.}\ (2\pi)^{-3/2} \int \overline{\phi(x,k)}\ f(x)\ d^3x$ *exists.* (l.i.m. *means* L^2*-limit of* $\int_{\{x|\ |x|<R\}}$).

(ii) $(E_{a.c.}f)(x) = (2\pi)^{-3/2}\ \text{l.i.m.} \int \phi(x,k)\ f(k)\ d^3k$. *Let* $\{\phi_n\}_{n=1}^N$ *be a complete family of orthonormal eigenfunctions.*[16] (l.i.m. *here means* L^2*-limit as* $R \to \infty$ *and* $\varepsilon \downarrow 0$ $\{k|\ |k|<R$ *and* $|k-k_o|>\varepsilon$, *all* $|k_0|^2 \varepsilon \mathcal{E}\}$).

(iii) $\|f\|^2 = \int |\hat{f}(k)|^2\ d^3k + \sum_{n=1}^{N} \|f_n\|^2 + \|E_{sing}f\|^2$.

(iv) *If* $[\alpha,\beta] \subset \mathcal{R}$, *then*

$$\|E_{[\alpha,\beta]}f\|^2 = \int_{\alpha<k^2<\beta} d^3k\ |\hat{f}(k)|^2.$$

(v) $f \epsilon D(H)$ *if and only if* $\int |k^4|\ |\hat{f}(k)|^2\ d^3k < \infty$ *and* $E_{sing}\ f \epsilon D(H)$, *in which case*

$$(Hf)(x) = \text{l.i.m.} \int |k|^2\ \hat{f}(k)\ \phi(x,k)\ d^3k + \sum_{n=1}^{N} f_n E_n \phi_n(x) + (E_{sing}Hf)(x).$$

Remarks

1. (ii) and (v) tell us that from a formal point of view,

$$H\phi(\cdot,k) = |k|^2\ \phi(\cdot,k).$$

---

[16] $N = \infty$ only in the unlikely case that infinitely many positive energy bound states contribute.

We don't consider to what extent this is true in terms of the actual differential operators; we do however note that it is at least true in a distributional sense (once we prove that $\hat{\ }$ is surjective as we will in Chapter V).

2. Our proof will follow the ideas of Ikebe's paper. (Theorem IV.25 is analogous to his Theorem 5). For references on eigenfunction expansions, one may consult page 2 of his work.

3. (iv) implies that σ_{sing} lies in $\mathcal{E}$.

4. We will see with the choice of ϕ below (IV.16) $\hat{\ }$ and Ω^+ are related by $\widehat{\Omega^+ f} = \hat{f}_0$ where $\hat{f}_0$ is the ordinary Fourier transform. It will then follow that $\hat{\ }$ is surjective.

5. Until we prove that $\hat{\ }$ is surjective, (v) does not imply that $\widehat{Hf}(k) = |k|^2 \hat{f}(k)$ but only that $\widehat{Hf} - k^2 \hat{f} \in (\text{Ran}\ \hat{\ })^{\perp}$.

We will choose the ϕ's to be the formal solutions of a Lippman-Schwinger type equation.

$$\phi(x,k) = e^{ik\cdot x} - \frac{1}{4\pi} \int \frac{e^{i|k|\,|x-y|}}{|x-y|} V(y)\, \phi(y,k)\, d^3y. \tag{IV.16}$$

Since (IV.16) does not, a priori, have a solution, we multiply by $V_{\|}^{1/2}$ and obtain:

LEMMA IV.26. *Whenever* $|\vec{k}|^2 \notin \mathcal{E}$, *there is a unique* L^2*-function* $\psi(x,k)$ *obeying*

$$\psi(y,k) = V_{\|}^{1/2}(y)\, e^{ik\cdot y} - \frac{1}{4\pi} \int V_{\|}^{1/2}(y) \frac{e^{i|k|\,|x-y|}}{|x-y|} V^{1/2}(x)\, \psi(x,k)\, d^3x \tag{IV.17}$$

The function $\phi(x,k)$ *defined by*

$$\phi(x,k) = e^{ik\cdot x} - \frac{1}{4\pi} \int \frac{e^{i|k|\,|x-y|}}{|x-y|} V^{1/2}(y)\, \psi(y,k)\, d^3y \tag{IV.18}$$

exists for almost every x *and obeys* (IV.16). *For any* $W \in L^1 \cap R$, $W_{\|}^{1/2} \phi \in L^2$.

Proof. $V \in R$ implies that the kernel of the integral equation (IV.17) is Hilbert-Schmidt and $V \in L^1$ implies the inhomogeneous term is L^2. Thus, the Fredholm alternative and $|k|^2 \notin \mathcal{E}$ imply that (IV.17) has a unique solution. Pick a function $W(x) > 0$ for all x with $W \in L^1 \cap R$. Then $W^{1/2}(x) \frac{e^{i|k|\,|x-y|}}{|x-y|} V^{1/2}(y) \in L^2(R^3 \times R^3)$ (see Section I.2) and thus $e^{i|k|\,|x-y|}\,|x-y|^{-1}\, V^{1/2}(y) \in L^2(y)$ a.e., in x so the integral in (IV.18) exists a.e., and $V_{\|}^{1/2}\phi = \psi$ so (IV.16) also holds. It follows from (IV.18) that $W^{1/2}\phi \in L^2$ under the conditions $W \in L^1 \cap R$. ∎

The crucial idea behind Ikebe's (and hence our) proof is to relate the functions ϕ to the Fourier transform of the Green's function. Following Ikebe, for $\mathrm{Im}\,\kappa > 0$. $\mathrm{Im}\,\kappa^2 \neq 0$, we define:

$$H(x,y,\kappa) \equiv G(x,y;\kappa^2). \qquad \text{(IV.19)}$$

By Theorems II.38 and II.41, $H(x,\cdot;\kappa) \in L^1(\cdot) \cap L^2(\cdot)$ a.e., in x. Thus, we can define

$$g(x,k;\kappa) = (2\pi)^{-3/2} \int H(x,y;\kappa)\, e^{ik\cdot y}\, d^3y \qquad \text{(IV.20)}$$

and

$$h(x,k;\kappa) = (2\pi)^{3/2} (|k|^2 - \kappa^2)\, g(x,k;\kappa). \qquad \text{(IV.21)}$$

LEMMA IV.27 (analogous to Ikebe's Lemma 9.2). *$h(x, k; \kappa)$ can be written in the form*

$$h(x,k;\kappa) = e^{ik\cdot x} - \frac{1}{4\pi} \int \frac{e^{i\kappa|x-y|}}{|x-y|} V^{1/2}(y)\, p(y,k;\kappa)\, d^3y \qquad \text{(IV.22)}$$

where p is the unique solution of

$$p(y,k;\kappa) = V_{\|}^{1/2}(y)\, e^{ik\cdot y} - \frac{1}{4\pi} \int V_{\|}^{1/2}(y) \frac{e^{i\kappa|x-y|}}{|x-y|} V^{1/2}(x)\, p(x,k;\kappa)\, d^3x. \qquad \text{(IV.23)}$$

Proof. According to Theorem II.40, $H(x,y;\kappa)$ obeys the integral equation:

$$H(x,y;\kappa) = \frac{e^{i\kappa|x-y|}}{4\pi|x-y|} - \frac{1}{4\pi}\int \frac{e^{i\kappa|x-z|}}{|x-y|} V(z)\, H(z,y;\kappa)\, d^3z. \qquad \text{(IV.24)}$$

If we take Fourier transforms with respect to y, temporarily deferring justification of an interchange of integration, we see that (Ikebe's (9.4)):

$$g(x,k;\kappa) = (2\pi)^{-3/2}\frac{e^{ik\cdot x}}{|k|^2-\kappa^2} - \frac{1}{4\pi}\int \frac{e^{i\kappa|x-y|}}{|x-y|} V(y)\, g(y,k;\kappa)\, d^3y$$

$$h(x,k,\kappa) = e^{ik\cdot x} - (4\pi)^{-1}\int |x-y|^{-1} e^{i\kappa|x-y|}\, V(y)\, h(y,k,\kappa)\, d^3y. \qquad \text{(IV.25)}$$

We obtain (IV.22) and (IV.23) from (IV.25) by a method analogous to that of Lemma IV.26. (That $V_{\|}^{1/2} h(\cdot,k,\kappa) \epsilon L^2(\cdot)$ follows from (IV.26) below).

To justify the interchange of integrals in taking the Fourier transform of (IV.24), we must show that

$$\int \frac{e^{-\text{Im}\kappa|x-z|}}{|x-z|} |V(z)|\; |H(z,y;\kappa)|\; d^3z\; d^3y < \infty$$

a.e., in x. Since $e^{-\text{Im}\kappa|x-z|}\,|x-z|^{-1}\,|V(z)|^{1/2} \epsilon L^2(z)$ a.e., in x, it is sufficient to show

$$\int d^3y\; |V(z)|^{1/2}\; |H(z,y;\kappa)| \;\epsilon\; L^2(z)$$

or that:

$$\int d^3y\; d^3x\; d^2z\; |V(z)|\; |H(z,y;\kappa)|\; |H(z,x;\kappa)| < \infty. \qquad \text{(IV.26)}$$

(IV.26) can be proven by the straight-forward but tedious method of plugging the integral equation (IV.24) in for H and using the fact that $V_{\|}^{1/2} G\, V_{\|}^{1/2}$ is Hilbert-Schmidt. Explicitly the integral in (IV.26) is (using (IV.24) and the symmetry of G bounded by the sum of four terms:

$$(I) = \int d^3y\, d^3x\, d^3z\, |V(z)|\, |H_0(z,y;\kappa)|\, |H_0(z,x;\kappa)|$$

$$(II)+(III) = 2\int d^3y\, d^3z\, d^3x\, d^3w\, |V(z)|\, |H_0(z,y;\kappa)|\, |H(z,w;\kappa)|\, |V(w)|\, |H_0(w,x;\kappa)|$$

$$(IV) = \int d^3y\, d^3z\, d^3x\, d^3w\, d^3s\, |V(z)|\, |H(z,w;\kappa)|\, |V(w)|\, |H_0(w,x;\kappa)|\, |H(z,s;\kappa)|\, |V(s)|\, |H_0(s,y;\kappa)|.$$

Since $H_0(x,y;\kappa)$ is an L^1-function of $|x-y|$, we immediately are able to do two integrations in the integrals (I), (II) + (III), (IV). We thus, need only show

$$(V) = \int d^3z\, |V(z)|$$

$$(VI) = \int d^3z\, d^3w\, |V(z)|\, |H(z,w;\kappa)|\, |V(w)|$$

$$(VII) = \int d^3\, d^3w\, d^3s\, |V(w)|\, |H(z,w;\kappa)|\, |V(z)|\, |H(z,s;\kappa)|\, |V(s)|$$

are finite. Since $V \epsilon L^1$, (V) is finite. Since $V_{||}^{1/2} G\, V_{||}^{1/2}$ is Hilbert-Schmidt (by Corollary II.35) (VI) and (VII) are of the form

$$\int |\psi(x)|\, |k(x,y)|\, |\psi(y)|\, d^3x\, d^3y$$

with $\psi \epsilon L^2$ and k Hilbert-Schmidt, and thus they are finite. As a result (IV.26) and with it the lemma is proven. ∎

LEMMA IV.28. *Let* $f \epsilon C_0^\infty$. *Then:*

$$\Phi(k;\kappa) = (2\pi)^{-3/2} \int \overline{h(x,k;\kappa)}\, f(x)\, d^3x \qquad \text{(IV.27)}$$

and

$$\hat{f}(k) = (2\pi)^{-3/2} \int \phi(x,k)\, f(x)\, d^3x \qquad \text{(IV.28)}$$

exist (i.e., the integrals are absolutely integrable). Let $[\alpha,\beta] \subset \mathcal{R}$. *Then* $\Phi(k;\kappa)$ *has an extension to* κ *real for* $\alpha^{1/2} < \kappa < \beta^{1/2}$ *with* $f(k) = \Phi(k,|k|)$ *(any* $|k|^2 \epsilon [\alpha,\beta]$*). Moreover* Φ *(extended) is uniformly continuous in* k *and* κ *in the region of all* k *and* $\alpha^{1/2} \leq \text{Re}\,\kappa \leq \beta^{1/2}$, $\text{Im}\,\kappa \geq 0$.

Proof. The functions $\tilde{V}(k,y) = V_{\|}^{1/2}(y)\, e^{ik\cdot y}$ are all in L^2. Moreover, by the dominated convergence theorem, $\tilde{V}(k,\cdot)$ is continuous in k as an L^2-valued function and since $\|V(k,\cdot) - V(k',\cdot)\|_2 = \|V(k-k',\cdot) - V(0,\cdot)\|_2$ it is continuous uniformly in k. Consider the kernel

$$R_\kappa = -(4\pi)^{-1}\, V_{\|}^{1/2}(y)\, e^{i\kappa|x-y|}\, |x-y|^{-1}\, V^{1/2}(x).$$

It defines a Hilbert-Schmidt kernel in the strip which is continuous and has $\|R_\kappa\| \to 0$ as $\operatorname{Im} \kappa \to \infty$. Since $[\alpha,\beta]$ contains no exceptional points, $(1 + R_\kappa)^{-1}$ exists for $\alpha^{1/2} < \kappa < \beta^{1/2}$ and by Corollary II.4, it exists when $\operatorname{Im} \kappa > 0$. Thus $(1 + R_\kappa)^{-1}$ is a uniformly norm continuous operator in the strip (since it is continuous on the compact set strip $\cup \{\infty\}$ with $(1-R_\infty)^{-1} = 1$). As a result $p(\cdot,k,x)$ $[= (1 - R_\kappa)^{-1}\, \tilde{V}(k,\cdot)]$ defined by (IV.23) can be continued to a uniformly continuous L^2-valued function defined in the whole strip. Moreover, $p(\cdot, k, |k|) = \psi(\cdot,k)$ in the notation of Lemma IV.26 (for $k^2 \epsilon [\alpha,\beta]$).

By (IV.22), $|f(\cdot)|^{1/2}\, h(\cdot, k; \kappa)$ defines an L^2-valued uniformly continuous and bounded function of k and κ in the strip)for $f \epsilon R$ implies $|f|^{1/2} G_0 V^{1/2}$ is a uniformly continuous in k kernel). Thus

$$\Phi(k;\kappa) = \int \frac{f(x)}{|f(x)|^{1/2}}\, [|f(x)|^{1/2}\, h(x,k;\kappa)]\, d^3x$$

is well-defined and uniformly continuous in the strip. Moreover,

$$p(y, k, |k|) = \psi(y,k) \text{ implies } \hat{f}(k) = \Phi(k,|k|). \blacksquare$$

The remainder of the proof of Theorem IV.15 is essentially a word-by-word translation of pp. 23-26 of Ikebe's paper. For the reader's convenience (and because of the minor differences in the proof), we sketch this beautiful argument, referring to Ikebe for various technical details.

LEMMA IV.29. *Let* $f \epsilon C_0^\infty$. *Let* $[\alpha,\beta] \subset \mathfrak{R}$. *Then:*

$$\|E_{[\alpha,\beta]} f\|^2 = \int_{\sqrt{\alpha}<|k|<\sqrt{\beta}} |f(k)|^2 \, d^3k.$$

Proof (following Ikebe). We first note that Parseval's equality for ordinary Fourier transforms implies that:

$$\int H(z,x;\kappa)\,\overline{H(z,y;\kappa)}\,dz = \int g(x,k;\kappa)\,\overline{g(y,k;\kappa)}\,dk$$

a.e., in x and y and for any κ with $\text{Im}\,\kappa > 0$, $\text{Im}\,\kappa^2 \neq 0$.

$$(\kappa^2-\bar{\kappa}^2) \int \overline{H(z,x;\bar{\kappa})}\,H(z,y,\bar{\kappa})\,dz$$
$$= \int \frac{2i\varepsilon}{(k^2-\mu)^2+\varepsilon^2}\,h(x,k;\kappa)\,\overline{h(y,k;\kappa)}\,\frac{dk}{(2\pi)^3} \tag{IV.29}$$

where $\kappa^2 = \mu + i\varepsilon$.

If we multiply the left hand side of (IV.29) by $\overline{f(x)}\,f(y)$ and integrate, we obtain:

$$(\kappa-\bar{\kappa}^2)\,\langle R_{\bar{\kappa}^2} f, R_{\bar{\kappa}^2} f\rangle = (\kappa^2-\bar{\kappa}^2)\,\langle f, R_{\kappa^2} R_{\bar{\kappa}^2} f\rangle = \langle f,(R_{\kappa^2}-R_{\bar{\kappa}^2})f\rangle \tag{IV.30}$$

where $R_\lambda = (H-\lambda)^{-1}$. To justify the interchange of integrals, we must show that

$$\int |f(x)|\,|H(z,x;\kappa)|\,|\overline{H(z,y;\kappa)}|\,|f(y)|\,dxdydz < \infty.$$

This follows from Corollary II.36, which implies that $|f(x)|^{1/2} H(z,x;\kappa)$ and $H(z,y;\kappa)\,|f(y)|^{1/2}$ are Hilbert-Schmidt kernels.

Multiplying the right side of (IV.29) by the same factors yields

$$\int dk\,\frac{2i\varepsilon}{(|k|^2-\mu)^2+\varepsilon^2}\,|\Phi(k;\sqrt{\mu+i\varepsilon})|^2. \tag{IV.31}$$

The interchange of integrals needed to obtain (IV.31) is allowed because $\int dx\,|f(x)|\,|h(x,k;\kappa)| < c$, independent of k (for fixed κ) from the proof of Lemma IV.28 and because $[(|k|^2-\mu)^2+\varepsilon^2]^{-1} \epsilon L^1(k)$.

Before putting (IV.29) - (IV.31) together, we remark that the functions

$$F_\varepsilon(x) = \int_\alpha^\beta \frac{1}{x-\mu-i\varepsilon} - \frac{1}{x-\mu+i\varepsilon}\, d\mu$$

$$= \int_\alpha^\beta \frac{2i}{(x-\mu)^2+\varepsilon^2}$$

obey $\frac{1}{2\pi i} F_\varepsilon(x) \to \begin{cases} 1 & \alpha<x<\beta \\ \tfrac{1}{2} & x=\alpha \text{ or } \beta \\ 0 & x\notin[\alpha,\beta]. \end{cases}$

Thus, by the functional calculus, and the fact that we have seen that α and β are not eigenvalues,[17] one has:[18]

$$\langle f, E_{[\alpha,\beta]} f\rangle = \frac{1}{2\pi i} \lim_{\varepsilon\downarrow 0} \int_\alpha^\beta \langle f, (R_{\mu+i\varepsilon} - R_{\mu-i\varepsilon}) f\rangle \, d\mu. \qquad \text{(IV.32)}$$

Putting (IV.29) - (IV.32) together, we obtain:

$$\|E_{[\alpha,\beta]} f\|^2 = \lim_{\varepsilon\downarrow 0} \frac{1}{\pi} \int_\alpha^\beta d\mu \int dk \, \frac{\varepsilon}{(|k|)^2-\mu)^2+\varepsilon^2} \, |\Phi(k;\sqrt{\mu+i\varepsilon})|^2.$$

Three steps now complete the proof:

(1) Since $|\Phi(k;\sqrt{\mu+i\ })|$ is uniformly bounded for all k and $\alpha<\mu<\beta$, we can interchange the μ and k integrations.

(2) By a short computation (see Ikebe, p. 24), we can justify taking $\lim_{\varepsilon\downarrow 0}$ inside the dk integral.

(3) By using the standard fact for g continuous (see Appendix 2):

$$\lim_{\varepsilon\downarrow 0} \pi \int_\alpha^\beta d\mu \, \frac{\varepsilon}{(x-\mu)^2+\varepsilon^2} \, g(x) = \begin{cases} g(\mu) & \alpha<\mu<\beta \\ 0 & \mu\notin[\alpha,\beta] \end{cases}$$

and the fact that $\Phi(k,|k|) = \hat{f}(k)$, we finally obtain Lemma IV.29. ∎

---

17 Any positive eigenvalue is an exceptional point by Theorem III.6.

18 This result goes back at least as far as Stone [T26], p. 183.

It is now reasonably straightforward to complete the proof of Theorem IV.15:

(1) Since $\mathcal{R}$ is open, we can write $\mathcal{R} = \bigcup_{i=1}^{\infty} [a_i,a_{i+1}]$ where the sets intersect only in the end points.

(2) Since the endpoints are in $\mathcal{R}$, they cannot be eigenvalues, so

$E_{\mathcal{R}} = \text{s-lim} \sum_{i=1}^{\infty} E_{[a_i,a_{i+1}]}$: Thus, for $f \in C_0^{\infty}$, we have

$$\|E_{\mathcal{R}}f\|^2 = \int |\hat{f}(k)|^2 \, d^3k.$$

(3) Given any $f \in L^2$, let $f_n \in C_0^{\infty}$ be chosen so that for $f_n \overset{L^2}{\to} f$. Then $E_{\mathcal{R}}f_n \overset{L^2}{\to} E_{\mathcal{R}}f$. On the other hand $\int |\hat{f}_n - \hat{f}_m|^2 \, d^3k = \|E_{\mathcal{R}}(f_n - f_m)\|^2 \to 0$ so $\hat{f}_n(k)$ is L^2 Cauchy and so converges to a function $\hat{f}(k)$ with

$$\|E_{\mathcal{R}}f\|^2 = \int |\hat{f}(k)|^2 \, dk$$

and similarly

$$\int_{a<|k|^2<\beta} |\hat{f}(k)|^2 \, d^3k = \|E_{[a,\beta]}f\|^2$$

if $[a,\beta] \subset \mathcal{R}$.

(4) $\hat{f}(k)$ defined in (3) is independent of the approximating sequence since $\int |(\hat{f}-\hat{g})(k)|^2 \, d^3k = \|E_{\mathcal{R}}(f-g)\|^2$. Thus $f_n \overset{L^2}{\to} f \Longrightarrow \hat{f}_n \overset{L^2}{\to} \hat{f}$.

(t) If f vanishes outside $\{x \mid |\vec{x}| < R\} \equiv B$, then

$$\hat{f}(k) = (2\pi)^{-3/2} \int \overline{\phi(x,k)} \, f(x) \, dx.$$

For $\phi(x,k)\, \chi_B(x) \in L^2$ by (IV.18), so the integral exists and $f_n \overset{L^2}{\to} f (f_n \in C_0^{\infty})$ implies $\hat{f}_n(k) \to (2\pi)^{-3/2} \int \overline{\phi(x,k)} \, f(x) \, d^3x$.

(6) Combining (4) and (5), we see that

$$f(k) = \text{l.i.m.} \; (2\pi)^{-3/2} \int \phi(x,k) \, f(x) \, d^3x$$

so that (i) and (iv) of the theorem are proven.

(7) From (3), Ran $E_{\mathfrak{R}} \subset \mathcal{H}_{a.c.}$ for $\|E_\Omega E_{\mathfrak{R}} f\|^2 = \int_\Omega |f(k)|^2\, d^3k$ while in a spectral representation

$$\|E_\Omega E_{\mathfrak{R}} f\|^2 = \int_\Omega |(E_{\mathfrak{R}} f)_{a.c.}(\lambda)|^2\, d\mu_{a.c.}(\lambda) + \int_\Omega |(E_{\mathfrak{R}} f)_{non\text{-}a.c.}(\lambda)|^2\, d\mu_{non\text{-}a.c.}(\lambda)$$

so that $|(E_{\mathfrak{R}} f)_{non\text{-}a.c.}(\lambda)|^2\, d\mu_{non\text{-}a.c.}(\lambda)$ is an absolutely continuous measure which is only possible if $(E_{\mathfrak{R}} f)_{non\text{-}a.c.} = 0$. Moreover since $\mathcal{E}$ has measure 0, $E_{\mathcal{E}} E_{a.c.} = 0$ and since the spectrum is discrete in $(-\infty, 0)$, $E_{(-\infty,0)} E_{a.c.} = 0$ so $E_{\mathfrak{R}-\mathfrak{R}} E_{a.c.} = 0$. Thus $E_{\mathfrak{R}} \equiv E_{a.c.}$. This proves (iii) of the theorem.

(8) To prove

$$(E_{a.c.} f)(x) = \text{l.i.m.}\ (2\pi)^{-3/2} \int \phi(x,k)\, f(k)\, d^3k$$

we need only show

$$(E_{[\alpha,\beta]} f)(x) = (2\pi)^{-3/2} \int_{\sqrt{\alpha}<|k|<\sqrt{\beta}} \phi\,(x,k)\, \hat{f}(k)\, d^3k \qquad \text{(IV.33)}$$

whenever $[\alpha,\beta] \subset \mathfrak{R}$ for $E_{\mathfrak{R}} = E_{a.c.}$

(9) To prove (IV.33), let $g \in C_0^\infty$. Since

$$\|E_{[\alpha,\beta]} f\|^2 = \int_{\sqrt{\alpha}<|k|<\sqrt{\beta}} |\hat{f}(k)|^2\, d^3k,$$

it follows by polarization that

$$\begin{aligned} \langle g, E_{[\alpha,\beta]} f\rangle &= \int_{\sqrt{\alpha}<|k|<\sqrt{\beta}} \hat{f}(k)\, \hat{g}(k)\, d^3k \\ &= (2\pi)^{-3/2} \int_{\sqrt{\alpha}<|k|<\sqrt{\beta}} \hat{f}(k) \int \phi(x,k)\, g(x)\, d^3x\, d^3k \\ &= (2\pi)^{-3/2} \int d^3x\, \left[\int \phi(x,k)\, \hat{f}(k)\, dk \right] g(x). \qquad \text{(IV.34)} \end{aligned}$$

The interchange of the order of integration is justified since

$$\int |\phi(x,k)|\ |g(x)|\ dx$$

is bounded for $\sqrt{\alpha} < |k| < \sqrt{\beta}$ by (IV.18) and $\hat{f}(k)$ is L^1 over the same region.

(10) Since $\langle g, E_{[\alpha,\beta]}f\rangle = \int_{\sqrt{\alpha}<|k|<\sqrt{\beta}} d^3k\ \hat{f}(k)\ \overline{\hat{g}(k)}$ we have that $\langle g, E_{[\alpha,\beta]}Hf\rangle = \int_{\sqrt{\alpha}<|k|<\sqrt{\beta}} d^3k\ |k|^2\ \hat{f}(k)\ \overline{\hat{g}(k)}$ so as in (8)-- (9):

$$(E_{a.c.}Hf)(x) = (2\pi)^{-3/2}\ \text{l.i.m.} \int |k|^2\ \hat{f}(k)\ \phi(x,k)\ d^3k$$

from which (v) of the theorem follows. This completes the proof. ∎

We are now in a position to understand why the approximation theorems of Section IV.4 are valid and why they depend on $V_n \to V$ in L^1 and *R-norms*. As we shall see in Chapter V, there is a very close connection between the wave operators and the eigenfunction expansion [in light of the Lippmann-Schwinger form of (IV.16), this is to be expected]. In fact $\Omega_n^{\pm}$ will converge to $\Omega^{\pm}$ when $V_n \xrightarrow{L^1} V$ and $\psi_n(\cdot,k) \to \psi(\cdot,k)$ [in $L^2(\cdot)$] where ψ_n solves (IV.17) with V replaced by V_n. It is now clear why $V_n \xrightarrow{R} V$ and $V_n \xrightarrow{L^L} V$ imply convergence of $\Omega^{\pm}$; for $V_n \xrightarrow{L^1} V$ tells us the homogeneous terms of IV.7 converge in L^2 and $V_n \xrightarrow{R} V$ implies (following our discussion in Remark 1 following Theorem IV.23) the kernels of the integral equations converge in Hilbert-Schmidt norm. It is also clear that R-boundedness should not be enough.

Appendix 1 to IV.5. *Solutions of Hilbert-Schmidt integral equations with kernels continuous on the boundary of a region.*

What we have chosen to call the analytic Fredholm theorem (Theorem A.27) tells us that a compact operator-valued function analytic in a region D with $[1 - f(\lambda_o)]$ invertable for some λ_0 has $(1-f(\lambda))^{-1}$ meromorphic in D. For the kernel $R_{|k|}$ with $|k|$ real, we are in the situation of $f(\lambda)$

having a continuous extension to the region $\bar{D}$; in this case, we want to know about the existence of $(1-f(\lambda))^{-1}$ for $\lambda\epsilon\partial D$. The crucial fact from "classical" complex analysis is:

THEOREM IV.30. *A non-zero function which is analytic in the open unit disc and continuous on the closed unit disc, has the property that* $\{z\,|\,|z|=1,\ f(z)=0\}$ *has zero (linear) Lebesgue measure in the unit circle.*[19]

Proof. See Hoffman [T15], pp. 52; 77. ∎

Remark

By a simple conformal mapping, the theorem holds for a function analytic in the interior of a semi-circle, continuous on the boundary.

THEOREM IV.31. *Let* $f(\lambda)$ *be a compact-operator valued function analytic in the upper half plane with* $(1-f(\lambda_0))$ *invertable for some* λ_0. *Let* $f(\lambda)$ *have an extension to a function continuous in the closed upper half-plane. Then* $S=\{\lambda|\lambda$ *is real and* $(1-f(\lambda))^{-1}$ *does not exist*$\}$ *is closed and has measure zero (in* $\mathbb{R}$).

Proof. It is enough to show for any λ that $S\cap[\lambda-\varepsilon,\lambda+\varepsilon]$ is closed of measure 0 for some ε (then one merely invokes σ-compactness of the real line). By the continuity assumption given λ, we can pick ε, so that $|\mu-\lambda|<\varepsilon$; Im $\mu\geq 0$ implies $\|f(\mu)-f(\lambda)\|<\frac{1}{3}$. By the compactness of $f(\lambda)$, we can pick $A=\sum_{n=1}^{N}|\psi_n\rangle\langle\phi_n|$ of finite rank, so that $\|f(\lambda)-A\|<\frac{1}{2}$, so $(1-f(\mu))=\{1-A[1-f(\mu)+A]^{-1}\}[1-f(\mu)+A]$. Thus $(1-f(\mu))^{-1}$ exists if and only if $[1-A(1-f(\mu)+A)^{-1}]^{-1}$ exists. A simple computation (see e.g., Hunziker [54], p. 461) shows that this is so if and only if $(1-a(\mu))^{-1}$ exists where:

---

[19] Moreover, it has been proven by Fatou [31], that any closed set of measure zero on the circle is a suitable candidate for the set of zeros of some regular function. See, e.g., Hoffman [T15], pp. 80-81.

$a(\mu) = A(1 - f(\mu) + A)^{-1}$ restricted to $[\psi_1, \ldots, \psi_N]$. $(1 - a(\mu))^{-1}$ exists if and only if $\det(1 - a(\mu)) \neq 0$. But $a(\mu)$ is continuous on the semicircle, analytic in the interior so $\det(1 - a(\mu))$ has these properties. Moreover, this det does not vanish identically, for then (by a continuation argument) $(1 - f(\lambda_0))^{-1}$ couldn't exist. As a result, we can apply Theorem IV.30 which implies $[\lambda - \varepsilon, \lambda + \varepsilon] \cap S$ has measure zero. ∎

COROLLARY IV.32. *$\mathfrak{E}$ is closed of Lebesgue measure* 0 ($\equiv$ *Lemma IV.24*). ∎

Remark

Theorem IV.21 extends to any closed region, D, where ∂D obeys a regularity condition, that D is homeomorphic to a disc by a homeomorphism conformal on the interior.

Appendix 2 to IV.5. A δ-function computation.

Since the standard proofs that $\lim_{\varepsilon \downarrow 0} \frac{\varepsilon}{x^2 + \varepsilon^2} = \pi\delta(x)$ are either purely formal or use the full C^∞ nature (at least C^1!) of the test functions, we present (for the reader's convenience) a short rigorous derivation of:

LEMMA IV.23. *Let* $f(x)$ *be continuous at* $x = 0$; *and suppose* $f \in L^\infty$. *Then:*

$$\lim_{\varepsilon \downarrow 0} \int_{-\infty}^{\infty} \frac{\varepsilon}{x^2 + \varepsilon^2} f(x)\, dx = \pi f(0).$$

Proof. Letting $x = y\ $, we see that

$$\int_{-\infty}^{\infty} \frac{\varepsilon}{x^2 + \varepsilon^2} f(x)\, dx = \int_{-\infty}^{\infty} \frac{\ }{y^2 + 1} f(y\ )\, dy$$

$$= \pi f(0) + \int_{-\infty}^{\infty} \frac{1}{y^2 + 1} (f(y\varepsilon) - f(0))\, dy.$$

The integrand in the last integral converges pointwise to zero and is bounde by $2\|f\|_\infty (y^2 + 1)^{-1} \in L^1$ so the integral has 0 limit. ∎

Remark

In the proof of Lemma IV.29, we need

$$\lim_{\varepsilon \downarrow 0} \int_{\infty}^{\infty} \frac{\varepsilon}{x^2+\varepsilon^2} f(x;\varepsilon)\, dx = \pi f(0;0).$$

This follows from joint continuity in x and ε at zero and uniform boundedness in x and ε.

Bibliographic Note to Section IV.5

After completion of a preliminary draft of Chapter IV, I became aware of some recent work of Kuroda on eigenfunction expansions [80, 81, 82, 83].[20] Kuroda's work involves an approach of much greater generality and complexity than that which we have used. (We must emphasize once again that no new mathematical ideas have been added to Ikebe's work in what we have done in IV.5; only various formal simplifications have been achieved – and a different class has been studied. On the other hand, Kuroda's work involves new and powerful mathematical tools.) In one of the articles [81], Kuroda mentions in passing that he can use his method to treat potentials $V \in L^{3/2}$ and it is clear he only needs $V \in R$. However, it is not clear exactly what results he does obtain (unfortunately the article is in Japanese and Kuroda doesn't mention the $L^{3/2}$ potentials in his principal English language article [83]). In any event, it is likely his results improve (and predate!) Section IV.5. However, we must remark that the methods used above are conceptually simpler and thus probably more accessible to physicists than the more comprehensive Kuroda approach.

IV.6. *Asymptotic Completeness*

Since Theorem IV.11 tells us that $H_0 + V$ is (KC) when $V \in L^1 \cap \mathcal{R}$, we know (by Theorem IV.2) that $H_0 + V$ is (AC) if and only if $\mathcal{H}_{\text{sing}} = 0$. By

[20] See [80] for additional references.

Theorem IV.25, $\mathcal{H}_{\text{sing}} \subset \text{Ran}\, E_{\mathcal{E}}$ so $\sigma_{\text{sing}} \subset \mathcal{E}$. A non-zero singular spectrum must be uncountable.[21]

THEOREM IV.34. *A sufficient condition for* (AC) *to hold when* $V \in R \cap L^1$ *is that the exceptional set* $\mathcal{E}$ *be countable.* ∎

COROLLARY IV.35. *Let* $V \in L^1$ *and suppose* $\int \frac{|V(x)|\,|V(y)|}{|x-y|^2}\, d^3x\, d^3y < (4\pi)^2$. *Then* $H_0 + V$ *obeys* (AC); *in fact* $\Omega^{\pm}$ *are unitarily equivalences of* H_0 *and* H.

Proof. $\|V\|_R < 4\pi$ implies $\mathcal{E} = \phi$ and that there are no bound states of either positive or negative energy. Thus $\mathcal{H}_{\text{a.c.}} = \mathcal{H}$ so $\mathcal{H}_{\text{in}} = \mathcal{H}_{\text{out}} = \mathcal{H}$. ∎

COROLLARY IV.36. *Let* $V \in L^1 \cap R$ *of finite range (see Section I.5). Then* $H_0 + V$ *obeys* (AC).

Proof. In this case, $V_{\|}^{1/2}(x)\, \frac{e^{ik|x-y|}}{|x-y|}\, V^{1/2}(y)$ defines a compact operator for $\text{Im}\, k \geq -C$ $(C > 0)$. Thus $\mathcal{E}$ is discrete by the analytic Fredholm theorem and thus a finite set (since $\mathcal{E}$ is bounded by the Zemach-Klein theorem). ∎

Remarks

1. Using time-independent methods, Kato [70] has proven Corollary IV.35 without the assumption $V \in L^1$. Our proof tends to explain why such a result is true.

2. Our result Corollary IV.36 is much weaker in the asymptotic behavior allowed than Ikebe's $0(x^{-2-\varepsilon})$. However, our result requires no smoothness and does not require information about the absence of positive energy bound states as input.

3. Undoubtedly, stronger results about asymptotic completeness exist.

---

21 Remember that our σ_{sing} has all pure points removed.

Notes added in proof:

1. To prove (AC), one does not need the full eigenfunction expansion; it is enough to control $\lim_{\varepsilon \downarrow 0} < \psi, [(H-E+i\varepsilon)^{-1} - (H-E-i\varepsilon)-1]\psi>$ for a for a dense set of ψ. Thus Corollary II.35 holds without $V \epsilon L^1$ (see P. Rejto, J. Math. Anal. App. 17 (1967) 435 for a discussion of this point). Since finite range potentials are always in L^1 we cannot improve Corollary II.36, but we can simplify its proof with this remark.

2. Our result that σ_{sing} is a subset of some closed set of measure zero was also obtained independently in [74].

CHAPTER V

TIME-INDEPENDENT SCATTERING THEORY

V.1. *Introduction*

In Sections IV.1 through IV.4, we studied scattering theory by direct consideration of $e^{+iHt}\, e^{-iH_0 t}$. In contradistinction to such an approach, the "time-independent" method uses, as a starting point, formal equations for $\Omega^{\pm}$ and/or S, which no longer mention time explicitly. Actually, the method is used so differently in the hands of physicists and mathematicians that one is really dealing with two distinct programs when describing "time-independent scattering theory."

The mathematical approach goes back at least as far as the work of Friedrichs [38] and involves a three step procedure: (1) the "derivation" of a formal equation for some scattering theoretic object such as $\Omega^{\pm}$; (2) the rigorous proof of the solubility of the formal equation; (3) the proof that the solution of the equation is actually the scattering theory object. For an example of the method and additional references, see Kato [T17], pp. 551-565. These methods are generally extremely powerful and are behind the results of Kato [70] and Kato-Kuroda [74], which we have quoted in Chapter IV. Since we have proven a maximally good result on (WAC) without detailed consideration of these methods,[1] we will not study them in detail.

In some sense, the time-independent approach in physics goes back to the earliest days of quantum mechanics. In the more modern era, the term

[1] We have used them implicitly, of course, when we quote Kato and Kato-Kuroda.

is associated with using the Lippmann-Schwinger equation as a starting point. The physicist does not usually worry about rigorously establishing the connection between the Lippmann-Schwinger equation and the time-dependent equation but freely uses the L.-S. equation as a tool. In this sense, the spirits of the two approaches we have just described are opposites.

In this chapter, we stake out a middle ground. We are not interested in using time-independent methods as an existence tool, but would like to prove the standard physicists' formulae: the link between $\Omega^{\pm}$ and the Lippmann-Schwinger wave functions $\widehat{(\Omega^{+}f} = \hat{f}_0$ in the notation of IV.5, letting $\hat{f}_0$ be the ordinary Fourier transform of f), (Section V.4) and the usual form for the on-shell T-matrix (S-1) as $\langle\psi|V|\phi\rangle$ where ψ is a free state and ϕ an "interacting" state.

To my knowledge, the only previous direct proof of the Lippmann-Schwinger equation is due to Ikebe [56] for smooth potentials, although Hunziker [55] has sketched a proof whose details he claims should be tractable by the use of Ikebe's method. Originally, we had hoped to mimic Ikebe (which can be done!) but in studying his paper we found a circularity of argument which we describe in Section V.2. We then present, in Section V.4, a simple modification of his argument avoiding this circularity. For the reader's convenience, we first discuss Abelian limits in V.3. Finally, in Section V.5, we derive the usual formula for the on-shell T-matrix.

V.2. *The Circularity in Ikebe's Argument*

The argument that Ikebe uses to establish the connection between $\Omega^{\pm}$ and his eigenfunction transform is very beautiful. There seems however to be a subtle error. By propagation, this error is implicit in the work of Thöe [108] and Bertero et al., [134] since these authors base their arguments on Ikebe's paper.

The problem is the following: In Section IV.5 above and §1 - §9 of Ikebe's paper one proves that certain operators A have $\hat{\ }$-space kernels A(k) in the sense

$$\langle f, Ag\rangle = \int d^3k\ \overline{\hat{f}(k)}\ A(k)\hat{g}(k) \qquad (V.1)$$

when f or g is in $\mathcal{H}_{a.c.}$ so that:

$$\widehat{Ag}(k) - A(k)\hat{g}(k) \in (\text{Ran}\ \hat{\ })^{\perp}. \qquad (V.2)$$

Until it is established that Ran $\hat{\ }$ is dense, one cannot use "strong" formulae of the type $\widehat{Ag}(k) = A(k)\widehat{g(k)}$ or equivalently

$$\langle h, \widehat{Ag}\rangle = \int \overline{h(k)}\ A(k)\hat{g}(k)\ d^3k$$

for arbitrary h. We remark in passing that a result of the form

$$(Ag)(x) = (2\pi)^{-3/2}\int d^3k\ \phi(k, k)\ A(k)\hat{g}(k) \qquad (V.3)$$

is only *apparently* stronger than (V.1). It is merely a consequence of isometry $[(\hat{\ })^*(\hat{\ }) = 1]$ and (V.1) and is *not* directly related to the "strong" formula

$$\widehat{Ag}(k) = A(k)\hat{g}(k) \qquad (V.4)$$

which requires $\hat{\ }$ to be onto $[(\hat{\ })(\hat{\ })^* = 1]$. Only after the surjective nature of $\hat{\ }$ is established can formulae of form (V.4) be used, e.g.,

$$\widehat{Hf}(k) = |k|^2\,\widehat{f(k)} \qquad (V.5)$$

or

$$\widehat{E_{\Omega}f}(k) = X_{\Omega}(k)\hat{f}(k) \qquad (V.6)$$

or, most crucially:

$$\widehat{e^{iHt}f}(k) = e^{i|k|^2t}\hat{f}(k). \qquad (V.7)$$

The surjective nature is only established by Ikebe on page 31 of his paper, while on page 28, the culprit equation

$$\lim \int e^{ik\cdot x}\ \widehat{Hf}(k)\ d^3k = \lim \int e^{ik\cdot x}\ |k|^2\hat{f}(k)\ d^3k \qquad (V.8)$$

appears. This is equivalent to (V.5), which he has not yet proven.

And (V.8) is used crucially in the remainder of Ikebe's proof. It is used to prove his equation (11.6) which in turn is used in (11.8), the starting point of a lengthy computation.

We will see that Ikebe's proof can be modified to give a correct proof (Section V.4). In fact, the idea of the correction is simple. Both Ikebe and we are aiming for an equation:

$$(\Omega^{+}f)(x) = (2\pi)^{-3/2} \lim \int \phi(x, k) \hat{f}_0(k) \tag{V.9}$$

where $\hat{f}_0$ is the ordinary Fourier transform. Ikebe defines the right-hand side of (V.9) as $U_{-}f$ and proves:

$$(U_{-}^{*}f)(k) = (2\pi)^{-3/2} \lim \int e^{ik\cdot x}\hat{f}(k)\, dk \tag{V.10}$$

He then proves (and it is in this proof that the error appears) that $U_{-}^{*}\Omega^{+} = 1$. This proof requires formulae like $U_{-}^{*}e^{iHt} = e^{iH_0t}U_{-}^{*}$ and this in turn depends on the unproven (V.7). The way out is simple. It is to prove $U_{-} = \Omega^{+}$ directly. In that way formulae involving U_{-}^{*} never enter and we stay within the "eigenfunction" image.

Note added:

Prof. Ikebe has informed me (private communication) that the hole can be corrected using techniques in [144]. His method depends on smoothness of V and thus will not carry over to our case.

Note added in proof:

Professor Ikebe has also found a general proof (paper in preparation) that the weak formulae do, in the presence of other information, imply the the strong formulae. His method applies to both the smooth $0(r^{-2-\varepsilon})$ case and our $L^1 \cap R$ case.

V.3. *Abelian Limits*

The one common feature of all time independent equations is the occurrence of $\lim_{\varepsilon\downarrow 0}$. These invariably arise through the taking of an Abelian limit.

DEFINITION. We say that $\int_0^t f(s)\,ds$ has Abelian limit a, if

$$\lim_{\varepsilon\downarrow 0} \int_0^\infty e^{-\varepsilon s} f(s)\,ds = a.$$

This is a legitimate summability method in the sense that:

LEMMA V.1. *Let* f(s) *be a bounded measurable function and let* $\lim_{t\to\infty} \int_0^t f(s)\,ds = a$. *Then* $\lim_{\varepsilon\downarrow 0} \int_0^\infty e^{-\varepsilon s} f(s)\,ds = a$.

Proof.[2] Let $g(t) = \int_0^t f(s)\,ds$ and $\phi(\varepsilon) = \int_0^\infty e^{-\varepsilon s} f(s)\,ds$. Then g(t) is continuous, g(0) = 0; g(t) → a as t → ∞. Thus g(t) is bounded. Moreover $g'(t) = f(t)$ a.e. Thus

$$\begin{aligned}\phi(\varepsilon) &= \int_0^\infty ds\, e^{-\varepsilon s} f(s)\\ &= \lim_{t\to\infty} \int_0^t e^{-\varepsilon s} g'(s)\\ &= \lim_{t\to\infty} \left[\int_0^t \varepsilon e^{-\varepsilon s} g(s)\,ds + e^{-\varepsilon t} g(t)\right]\end{aligned}$$

$$\phi(\varepsilon) = \int_0^\infty \varepsilon e^{-\varepsilon s} g(s). \tag{V.10}$$

Thus $\phi(\varepsilon) - a = \int_0^\infty \varepsilon e^{-\varepsilon s} [g(s) - a]\,ds$. Given δ, pick T so large that $|g(s) - a| < \frac{1}{2}\,\delta$ for s > T. Then:

$$|\phi(\varepsilon) - a| \le \int_0^T \varepsilon e^{-\varepsilon s}\, |g(s) - a| + \frac{\delta}{2} \le \varepsilon T(a + \|g\|_\infty) + \frac{\delta}{2}.$$

Thus, we can pick ε so small that $|\phi(\varepsilon) - a| < \delta$. Since δ is arbitrary, the lemma is proven. ∎

A related result can be obtained based on the part of the proof of the last lemma that follows equation (V.10):

---

[2] A continuous analogue of the discrete case to be found in Hardy [T13], p. 72.

LEMMA V.2. *If* f(t) *obeys* $f(t) \to a$ *as* $t \to \infty$ *and is bounded and locally integrable, then*

$$\lim_{\varepsilon \downarrow 0} \varepsilon \int_0^\infty e^{-\varepsilon s} f(s) = a. \blacksquare$$

We thus have, using a vector-valued analogue of Lemma V.2, the Gell-Mann-Goldberger [41] definition of $\Omega^{\pm}$:

COROLLARY V.3.

$$\Omega^{\pm}\psi = \lim_{\varepsilon \downarrow 0} \int_0^{\mp\infty} dt \; \varepsilon e^{\pm \varepsilon t} \, e^{+iHt} \, e^{-iH_o t} \psi. \blacksquare$$

We can also take derivatives and obtain:

COROLLARY V.4. *Let* $V \in R$ *and let* $\psi, \phi \in \mathcal{H}_{+1}$. *Then*

$$\langle \psi, (\Omega^{\pm}-1)\phi \rangle = \lim_{\varepsilon \downarrow 0} \int_0^{\mp\infty} i \langle \psi, e^{+iHt} V e^{-iH_o t} \phi \rangle \, e^{\pm \varepsilon t} \, dt \qquad \text{(V.11)}$$

Proof. By Theorem II.10, under the hypothesis of the corollary

$$\frac{d}{dt} \langle \psi, e^{+iHt} e^{-iH_o t} \phi \rangle = i \langle \psi, e^{+iHt} V e^{-iH_o t} \phi \rangle,$$

so that

$$\langle \psi, (\Omega^{\pm}-1)\phi \rangle = \lim_{t \to \mp\infty} \int_0^t i \langle \psi, e^{+iHt} V e^{-iH_o t} \phi \rangle \, dt.$$

To finish, we need only show that the integrand is bounded. This follows from a computation using Theorems II.9 and II.10:

$$\begin{aligned} |\langle \psi, e^{+iHt} V e^{-iH_o t} \phi \rangle| &= |\langle e^{-iHt}\psi, V e^{-iH_o t} \phi \rangle| \\ &\le \|e^{-iHt}\psi\|_1 \, \|V e^{-iH_o t} \phi\|_{-1} \\ &\le \|e^{-iHt}\|_{1,1} \|\psi\|_1 \, \|\phi\|_1 \, \|V\|_{-1,1} \end{aligned}$$

is bounded. $\blacksquare$

(V.11) with its factor of $Ve^{-i(H_0 \pm i\varepsilon)t}$ looks suspiciously like a Lippmann-Schwinger type equation. In fact, it is the starting point for our proof of the connection of $\Omega^{\pm}$ and the Lippmann-Schwinger equation.

V.4. *Scattering and Eigenfunction Expansions*

Let us define the ordinary Fourier transform:

$$\hat{f}_0(k) = (2\pi)^{-3/2} \lim \int e^{-ik\cdot x} f(x)\, dx \tag{V.12}$$

The notation is quite natural.[3]

Given a "free" wave packet, f(x), the part of it that behaves like $e^{ik\cdot x}$ should go into $\phi(x, k)$ under Ω^+ since $\phi(x, k)$ is the solution of the Lippmann-Schwinger equation.[4]

Thus, we expect to find:

$$\widehat{\Omega^+ f} = \hat{f}_0. \tag{V.13}$$

As a first step toward proving (V.13), we prove:

THEOREM V.5. *Let* $V \in R \cap L^1$. *Then:*

$$\widehat{[(\Omega^+)^* f]}_0 = \hat{f}. \tag{V.14}$$

---

[3] For $\hat{f}_0$ is the eigenfunction transform associated with H_0, i.e., $V = 0$.

[4] *Formally* speaking, $\Omega^+ e^{ik\cdot x} = \phi(x, k)$ for

$$(\Omega^+)^* \phi = \phi - \int_0^{-\infty} ie^{+\varepsilon t} e^{+iH_0 t} Ve^{-iHt} \phi$$

(abelianizing)

$$= \phi - \int_0^{-\infty} ie^{i(H_0 - k^2 - i\varepsilon)t} V\phi$$

(for $H\phi = k^2\phi$ formally)

$$= \phi - \frac{1}{H_0 - k^2 - i\varepsilon} V\phi = e^{ik\cdot x}$$

so modulo the existence of bound states, $\Omega^+ e^{ik\cdot x} = \phi$. To see $H\phi = k^2\phi$ formally, we write

$$(H_0 - k^2)\phi = (H_0 - k^2)e^{ik\cdot x} + (H_0 - k^2)(H_0 - k^2 - i\varepsilon)^{-1} V\phi = V\phi.$$

That is, an interacting state, f, *whose eigenfunction transform is* $\hat{f}$, *has*

$$(e^{-iHt}f)(x) \to \int \hat{f}(k) \exp[ik\cdot x - ik^2 t]\, dk \text{ *weakly* as } t \to -\infty.$$

Proof. Let $M = \{f \in \mathcal{H}_{a.c.} | \hat{f}$ has compact support in some set

$$\{k | \alpha < |k|^2 < \beta\}, [\alpha, \beta] \cap \mathcal{E} = \phi\}.$$

We first prove:

$$<f, \Omega^+ g> = \int \overline{\hat{f}(k)}\, \hat{g}_0(k)\, d^3k \tag{V.15}$$

for $f \in M$ and $g \in C_0^\infty$ (x-space). For, by (V.11)

$$<f, \Omega^+ g> = <f, g> + \lim_{\varepsilon \downarrow 0} \int_o^{-\infty} <f, e^{+iHt} V e^{-iH_o t} g> e^{\varepsilon t}\, dt \tag{V.16}$$

But[5,6]

$$<f, e^{+iHt} V e^{-iH_o t} g> = \int d^3k\, \overline{\hat{f}(k)}\, e^{ik^2 t}\, \widehat{Ve^{-iH_o t} g}$$

$$= \int d^3k\, \overline{\hat{f}(k)}\, e^{ik^2 t} \int d^3x\, \overline{\phi(x,k)} V(x) e^{-iH_o t} g(x).$$

No l.i.m. is needed for the x integral since $e^{-iH_o t} g(x)$ is bounded (in x) and $V_{||}^{1/2}$, $V^{1/2} \phi \in L^2(x)$. Thus

$$\int_0^{-\infty} e^{\varepsilon t} <f, e^{+iHt} V e^{-iH_o t} g> dt$$

$$= \int_o^{-\infty} dt \int d^3k \int d^3x\, \overline{\hat{f}(k)}\, \overline{\phi(x, k)} V(x) [e^{-it(H_o - k^2 + i\varepsilon)} g](x) \tag{V.17}$$

---

5 $<f, e^{+iHt} g> = \int d^3k\, \overline{\hat{f}(k)}\, e^{ik^2 t} \hat{g}(k)$ for $g \in \mathcal{H}_{a.c.}$ is a weak formula of type (V.1) and can be proven from

$$<f, E_{[\alpha,\beta]} g> = \int_{\alpha < |k|^2 < \beta} dk\, \overline{\hat{f}(k)}\, \hat{g}(k).$$

6 We use $f \in M \subset \mathcal{H}_{a.c.}$

The integrand is bounded by $|\hat{f}(k)|\ |\phi(x,k)V_{\|}^{1/2}(x)|\ e^{-\varepsilon|t|}V_{\|}^{1/2}(x)\,\|\hat{g}_0\|_1\ 2\pi^{-3/2}$. This is absolutely integrable in x, t, and k since $\hat{f}(k) \in L^2(k)$ and vanishes if $|k|^2 < \alpha$ or $|k|^2 > \beta$ and $\|V_{\|}^{1/2}\phi\|\ L^2(x)$ is bounded for $\alpha < |k|^2 < \beta$. Thus, we can rearrange the order of integration. Since $g \in C_0^\infty$;

$$\int_0^{-\infty} dt\,(e^{-it(H_0-k^2+i\varepsilon)}g)(x) = -i[(H_0-k^2+i\ )^{-1}g](x)$$

$$= \frac{-i}{4\pi}\int \frac{e^{-i|x-y|\sqrt{k^2-i\varepsilon}}}{|x-y|}\,g(y)\,d^3y \qquad \text{(V.18)}$$

Thus, using (V.17) and (V.18):

$$\int_0^\infty e^{\varepsilon t} \langle f, e^{+iHt}Ve^{-iH_0t}g\rangle$$

$$= \frac{-i}{4\pi}\int d^3k \int d^3x \int d^3y\ \overline{\hat{f}(k)}\ \overline{\phi(x,k)}\ V(x)\,\frac{e^{-i|x-y|\sqrt{k^2-i\varepsilon}}}{|x-y|}\,g(y) \qquad \text{(V.19)}$$

This last integrand is bounded, independent of ε, by

$$|\hat{f}(k)|\ |\phi(x,k)\,V_{\|}^{1/2}(x)|\ (g_{\|}^{1/2}(y))\ (V_{\|}^{1/2}(x)|x-y|^{-1}\ g_{\|}^{1/2}(y)).$$

This is absolutely integrable and thus we can plug (V.19) into (V.16) and take the limit inside the integral and freely rearrange integrals:

$$\langle f, \Omega^+g\rangle - \langle f, g\rangle = \int d^3k\ \overline{\hat{f}(k)} \int d^3y\ g(y)$$

$$\times\ \overline{\frac{1}{4\pi}\int d^3x\ \frac{e^{+ik|x-y|}}{|x-y|}\ V(x)\ \phi(x,k)}$$

$$= -\int d^3k\ \overline{\hat{f}(k)} \int d^3y\ g(y)\ [\overline{\phi(y,k)} - e^{-ik\cdot y}]$$

$$= -\int d^3k\ \overline{\hat{f}(k)}\ [\hat{g}(k) - \hat{g}_0(k)].$$

---

7 $|(e^{-itH_0}g)(x)| = (2\pi)^{-3/2}|\int e^{+ik\cdot x}e^{-itk^2}g_0(k)\,dk| \le (2\pi)^{-3/2}\,\|g_0\|_1.$

This proves (V.15), and thus (V.14) for any $f \in M$ since C_0^∞ is dense in L^2. Since M is dense in $\mathcal{H}_{a.c.}$, (V.14) holds for any $f \in \mathcal{H}_{a.c.}$ by continuity of $\hat{}$, $\hat{}_0$, and $(\Omega^+)^*$. Finally, if $f \in \mathcal{H}_{a.c.}$ both sides of (V.14) vanish. ∎

COROLLARY V.6. *Let* $V \in R \cap L^1$. *Then* $\hat{}$ *is onto* $L^2(k)$.

Proof. Since Ω^+ is an isometry, $(\Omega^+)^* \Omega^+ = 1$ so $(\Omega^+)^*$ is surjective. Thus $\{(\Omega^+)^* f | f \in L^2(x)\} = L^2(x)$. Since $\hat{}_0$ is surjective,

$$\widehat{\{(\Omega^+)^* f_0} | f \in L^2(x)\} = L^2(k).$$

Thus, by (V.14), $\hat{}$ is surjective. ∎

Remark

(V.14) also implies that $(\Omega^+)^* f \neq 0$ if $f \neq 0$ and $f \in \mathcal{H}_{a.c.}$ so that $\mathcal{H}_{in} = \mathcal{H}_{a.c.}$. Thus, in the $V \in L^1 \cap R$ case we have an independent proof of (KC).

THEOREM V.7.

(a) $\widehat{Hf}(k) = |k|^2 \hat{f}(k)$ if $f \in D(H)$.

(b) $\widehat{e^{iHt} f}(k) = e^{i|k|^2 t} \hat{f}(k)$

(c) $\widehat{E_\Omega f}(k) = X_\Omega(k) \hat{f}(k)$

where $X_\Omega(k) = 1$ *on* $\{k \,|\, |k|^2 \in \Omega\}$ *and* 0 *on the complement*

(d) $\widehat{\Omega^+ f}(k) = \hat{f}_0(k)$ (V.13)

Proof. (c) follows from the "weak" equation

$$\langle f, E_\Omega E_{a.c.} g\rangle = \int_{|k|^2 \in \Omega} \hat{f}(k)\, \hat{g}(k)\, d^3k$$

(which follows from (v) of Theorem IV.25) and the surjective nature of $\hat{}$.

(a), (b) and in fact $\widehat{F(H)f}(k) = F(|k|^2)\hat{f}(k)$ follow directly from (c) and the spectral functional calculus. Finally, (d) follows from Theorem V.5 via the computation

$$\widehat{\Omega^+ f} = \widehat{[(\Omega^+)^* \Omega^+ f]}_0 = \hat{f}_0$$

THEOREM V.8.

$$(2\pi)^{-3/2} \text{ l.i.m. } \int d^3x\, \phi(x,k)\, (\Omega^- f)(x) = \hat{f}_0(-k). \tag{V.20}$$

Proof.

$$\begin{aligned} \Omega^- f &= \lim_{t\to+\infty} e^{+iHt} e^{-iH_0 t} f \\ &= \lim_{t\to-\infty} \overline{e^{-iHt} e^{+iH_0 t} \bar{f}}^{\,8} \\ &= \overline{\Omega^+ \bar{f}} \end{aligned}$$

Thus

$$\begin{aligned} \overline{\widehat{\Omega^- f}(k)} &= \overline{\widehat{\Omega^+ \bar{f}}}(k) \\ &= \overline{\hat{\bar{f}}_0}(k) \\ &= (2\pi)^{-3/2} \text{ l.i.m. } \overline{\int e^{-ik\cdot x}\, \overline{f(x)}\, dx} \\ &= (2\pi)^{-3/2} \text{ l.i.m. } \int e^{+ik\cdot x} f(x)\, dx = \hat{f}_0(-k). \end{aligned}$$

Similarly, $\overline{\widehat{\Omega^- f}} = \int d^3x\, \phi(x, k)\, (\Omega^- f)(x)$. ∎

Remark

(V.20) is an intuitive direct consequence of time reversal invariance. This is not surprising since we used time reversal invariance explicitly in the proof when we wrote $\overline{e^{iHt} f} = e^{-iHt}\bar{f}$.

---

8 Since $\overline{Hf} = H\bar{f}$, we have $\overline{e^{iHt} f} = e^{-iHt}\bar{f}$.

V.5. *The S-matrix as a p-space Kernel*

In this section, we will derive[9] the fundamental link between the $\phi(x, k)$ and the scattering amplitude, out of which arise the Born expansion, dispersion relations, etc. Let us define (this agrees with the Goldberger-Watson convention [T9]):

DEFINITION.

$$T(k, k') = (2\pi)^{-3} \int e^{-ik\cdot x} V(x)\, \phi(x, k')\, dx \qquad \text{(V.21)}$$

$$= (2\pi)^{-3} \int e^{-ik\cdot x} V^{1/2}(x)\, \psi(x, k')\, dx \qquad \text{(V.22)}$$

defined whenever $k' \notin \mathcal{E}$ for all k.

PROPOSITION V.9. T is well-defined for $k' \notin \mathcal{E}^{1/2}$ and T is uniformly continuous in any region $|k'|^2 \in [\alpha, \beta]$, $k \in R^3$ when $[\alpha, \beta] \cap \mathcal{E} = \phi$.

Proof. For $|k'|^2 \in [\alpha, \beta]$, $\psi(x, k') \in L^2$ and uniformly L^2-continuous. Since $V^{1/2}(x) \in L^2$, $V^{1/2}(x)\, \psi(x, k') \in L^1$ and is uniformly L^1-continuous (in k'). Thus $T(k, k')$ is well-defined and uniformly continuous in k (and k'). ∎

The crucial fact is that the on-shell (i.e., $k^2 = k'^2$) T-matrix is the usual "T-matrix" (i.e., $S = 1 - 2\pi i\, T\delta(E_f - E_i)$ in formal language).

THEOREM V.10. *Let* $f, g \in \mathcal{S}$ *with* $\hat{f}_0, \hat{g}_0$ *having supports (respectively) in* $\{k \mid \alpha_i < |k|^2 < \beta_i\}$ $(i = 1, 2)$ *with* $[\alpha_i, \beta_i] \cap \mathcal{E} = \phi$. *Then:*

$$\langle f, (S-1)g\rangle = (-2\pi i) \int d^3k\, d^3k'\, \hat{f}_0(k)\, \hat{g}_0(k')\, T(k, k')\, \delta(k^2 - k'^2) \qquad \text{(V.23)}$$

---

[9] A general semi-rigorous approach of obtaining the Lippmann-Schwinger equation which related to ours may be found in the work of Coester [21].

Proof.

$$
\begin{aligned}
\langle f, (S-1)g\rangle &= \langle f, (\Omega^- - \Omega^+)^* \Omega^+ g\rangle \\
&= \langle (\Omega^- - \Omega^+) f, \Omega^+ g\rangle \\
&= \lim_{T\to\infty} \int_{-T}^{T} \langle e^{+iHt}(iV)\, e^{-iH_0 t} f, \Omega^+ g\rangle\, dt \\
&= \lim_{\varepsilon\downarrow 0} (-i) \int_{-\infty}^{\infty} e^{-\varepsilon|t|} \langle e^{+iHt} V e^{-iH_0 t} f, \Omega^+ g\rangle\, dt \\
&= \lim_{\varepsilon\downarrow 0} (-i) \int_{-\infty}^{\infty} dt\, e^{-\varepsilon|t|} \int dk'\, \overline{\widehat{e^{+iHt} V e^{-iH_0 t} f}(k')}\, \widehat{\Omega^+ g}(k') \qquad (V.24)
\end{aligned}
$$

where in the next to the last step we proceed as in the proof of Corollary V.4 and in the last step we use $\Omega^+ g \in \mathcal{H}_{a.c.}$ so the inner product comes from the eigenfunction expansion alone. By Theorem V.7, $\widehat{\Omega^+ g}(k') = \hat{g}_0(k')$, so we need only compute:

$$\widehat{e^{+iHt}\, V e^{-iH_0 t} f}(k') = e^{+i|k'|^2 t}\, \widehat{V e^{-iH_0 t} f}(k')$$

(by Theorem V.7)

$$= (2\pi)^{-3/2} \int dx\, e^{+i|k'|^2 t}\, V(x)\, (e^{-iH_0 t} f)(x)\, \overline{\phi(x, k')}$$

where l.i.m. is not needed since $V\phi \in L^1(x)$. Since $f \in \mathcal{S}$, we have

$$(e^{-iH_0 t} f)(x) = (2\pi)^{-3/2} \int dk\, e^{+ik\cdot x}\, e^{-i|k|^2 t}\, \hat{f}_0(k)$$

Thus the expression in (V.24) is equal to:

$$
\begin{aligned}
&(-i)(2\pi)^{-3} \int_{-\infty}^{\infty} dt \int dk'\, dx\, dk\, e^{i(|k|^2 - |k'|^2)t + \varepsilon|t|} \\
&\qquad \times \overline{(V(x)\, \phi(x, k')\, e^{-ik\cdot x})}\, \hat{f}_0(k)\, \hat{g}_0(k')
\end{aligned}
$$

The integrand is absolutely integrable (since $\phi V \in L^1(x)$ with bounded $\|\ \|_1$-norm as $|k'|^2$ runs over $[a_2, \beta_2]$). Thus we can rearrange the order of integration and explicitly "do" the t and x integrations:

$$\langle f, (S-1)g\rangle = \lim_{\varepsilon\downarrow 0} \int dk\, dk'\, T(k, k')(-i) \frac{2\varepsilon}{(|k|^2-|k'|^2)^2+\varepsilon^2} \hat{f}_o(k)\, \hat{g}_o(k').$$

Since Tfg is uniformly continuous in the region of integration, we can take the $\varepsilon\downarrow 0$ limit and obtain (V.23). ∎

We will not discuss in detail the sense in which (V.23) carries over to arbitrary $f, g \in L^2$ since (V.23) should be sufficient to establish any links between T and experimentally measured cross-sections.[10] We do remark that without any effort, (V.23) can be established for f, g with $\hat{f}_o$, $\hat{g}_o$ continuous (with the $[\alpha, \beta]$ supports) rather than C^∞ and that (V.23) can be carried over to any $f, g \in \mathcal{S}$ if the integral is interpreted in a principal parts sense, i.e., as $\lim_{\varepsilon\downarrow 0}$ of the integral over $\{k \mid |k-k'| < \varepsilon, \text{ all } |k'|^2 \in \mathfrak{S}\}$.

Added Note.

Computations similar to the above are contained in a paper of Ikebe [142]. He deals with V Holder continuous and $0(r^{-3-\varepsilon})$ but obtains (V.23) for all smearing functions ψ, ϕ.

---

10 There does not appear to be any fully comprehensive discussion of how scattering a beam of particles off a many particle target yields the simple two-body answer (in some no multi-scattering approximation) – it is almost inconceivable that such an argument would need a better form of (V.23) than one with smooth $\hat{f}_o$, $\hat{g}_o$.

CHAPTER VI

ANALYTIC SCATTERING THEORY

VI.1. *Introduction*

In this chapter, we discuss analyticity of the "scattering amplitude" $T(k, k')$ for the case where $V \in L^1 \cap R$. The material we present is fairly standard and thus this chapter is primarily discursive. We include it for the following reasons:

First, one original motivation for studying R was that $L^1 \cap R$ is the natural set of conditions for the validity of forward dispersion relations via an L^2 non-determinantal Fredholm theory approach (see Section VI.2). One contribution of this monograph is to justify from "first principles" the starting point for this standard argument. By combining suitable sections from Chapters II-VI, one can provide a complete derivation of forward dispersion relations for a "maximal" class[1] from first principles.

Secondly, we wish to mention what consequences the exceptional set $\mathfrak{E}$ would have in the pathological case that it is non-empty. We will find that if V has finite range, $\mathfrak{E}$ decouples from scattering in the sense that any "singularities" at $\mathfrak{E}$ are removable singularities.

---

[1] The first Born approximation to the forward amplitude is not finite if $V \notin L^1$. Thus $V \in L^1$ is more-or-less needed for forward amplitudes to exist; $V \in R$ is almost necessary for the Hamiltonian to exist as a form, at least if V is permitted to have both signs. For more singular *positive* potentials, the Friedrichs extension can define the Hamiltonian. Martin [135] has shown how to prove once-subtracted forward dispersion relations for repulsive central potentials less singular than r^{-3} at $r = 0$ and of finite extent. And one can even prove the Mandelstam representation for potentials V which are real analytic, have exponential tails and singularities as bad as r^{-2} (ln r) at $r = 0$ [136]. Thus, *if* restrictions are placed on the sign of V, $L^1 \cap R$ is not even approximately maximal.

Finally, we wish to present a simple and mathematically complete treatment of this question. This treatment resembles that given by Tiktopoulos [110; unpublished], but we have added a few mathematical details (esp. Theorem VI.2).

The earliest proof of dispersion relations in potential scattering is due to Khuri [75] who used determinantal Fredholm theory in the spirit of Jost and Pais [63]. There is a rather enormous literature on this subject and so we refer the reader to Newton's excellent bibliography [T24, esp. p. 296]. Our particular approach via the factorization $V = V^{1/2} V_{||}{}^{1/2}$ seems to have first appeared in the work of Grossman and Wu [44, 45], at least for potentials of finite range. Our treatment of non-forward amplitudes (Section VI.3) follows the work of Hunziker [52] quite closely.

The "usual" (i.e., Khuri) approach to dispersion relations requires $\int_0^\infty rV(r)dr < \infty$ and $\int_0^\infty r^2V(r)dr < \infty$, for the central potential $V(r)$. The second condition is $V \in L^1$ and the first is not unrelated to R; in fact:

THEOREM VI.1. *If* V *is non-nasty, in* L^1 *and* $\int_0^1 r\,dr \int d\Omega\,|V(\vec{r})| < \infty$, *then* $V \in R$.

Proof. As usual, we consider only the central monotone case and use the r_t notation of Theorem I.5. Since $\int_0^{r_t} r|V(r)|dr > \frac{1}{2} tr_t{}^2$, we see that $r_t < Dt^{-1/2}$ at least for t large. If we change the r integral in $\int_0^1 r|V(r)|dr$ to a Stieltjes integral over t, we see that

$$\int_{t=|V(1)|}^{\infty} tr_t\, dr_t < \infty . \tag{VI.1}$$

But, in terms of the distribution function $m_V(t)$ of I.1 (Appendix 2), $V \in L^{3/2}$ is equivalent to $\int_0^\infty t^{3/2}\, dm_V < \infty$. Since $V \in L^1$, we need only show $\int_{V(1)}^\infty t^{3/2}\, dm_V < \infty$. Since $m_V(t) = \frac{4\pi}{3} r_t{}^3$ we need only show $\int_{V(1)}^\infty t^{3/2}\, r_t{}^2\, dr_t < \infty$. Since $r_t < Dt^{-1/2}$, this follows from (VI.1). Thus $V \in L^{3/2}$ and thereby in R. ∎

Remarks

1. $\int_0^1 r|V(r)|dr < \infty$ is (in the monotone case) logarithmically weaker than even $V \in L^{3/2}$. For $V = r^{-2} (\ln r)^{-\alpha}$ has $\int_0^1 r\, V(r)dr < \infty$ only if $\alpha > 1$ (and $V \in L^{3/2}$ if $\alpha > \frac{2}{3}$ and in R if $\alpha > \frac{1}{2}$).

2. In general, i.e., for possibly nasty potentials, this theorem is probably false. For V central and L^1, $V \in R$ only if

$$-\int_0^1 dx \int_0^x dy\, |xV(x)|\, |yV(y)|\, \ln\left[\frac{x-y}{x+y}\right] < \infty$$

(doing the angular integral). Since the logarithmic factor varies between 0 and $-\infty$, this is *probably* not directly comparable to

$$\int_0^1 dx \int_0^x dy\, |xV(x)|\, |yV(y)| < \infty.$$

VI.2. *Analyticity of the Forward Amplitude: Forward Dispersion Relations*

For convenience, we suppose V is a central potential although the entire analysis goes through for any potential if we study $T(\vec{k}, \vec{k})$ for $\vec{k} = k\hat{e}$ with $\hat{e}$ fixed. We also suppose $V \in R \cap L^1$ (and follow Hunziker [52], Grossman-Wu [44, 45] and Tiktopoulos [110] as mentioned in the introduction. We study $T(\vec{k}, \vec{k}')$ $(k^2 = k'^2)$ which is linearly related to the "usual" $f(k, \theta)$ where $\cos\theta = \hat{k} \cdot \hat{k}'$.[2] By definition (Section V.5):

$$T(\vec{k}, \vec{k}) = (2\pi)^{-3} \int e^{-ik\cdot x}\, V^{1/2}(x)\, \psi(x, k)\, dx \qquad (V.22)$$

where ψ solves:

$$\psi(\vec{x}, \vec{k}) = V_{\|}^{1/2}(x)\, e^{ik\cdot x} - \frac{1}{4\pi} \int d^3y\, V_{\|}^{1/2}(x) \times \frac{e^{i|k|\,|x-y|}}{|x-y|}\, V^{1/2}(y)\, \psi(y, k) \qquad (IV.17)$$

---

[2] $f(k, \theta) = -2\pi^2\, T(k, k')$.

Following Hunziker, we let $\vec{k} = k\hat{e}$ with fixed $\hat{e}$ and define:

$$\chi(\vec{x}, k) = e^{-ik\hat{e}\cdot\vec{x}}\, \psi(\vec{x}, k\hat{e})$$

so that

$$T(k, \theta = 0) = (2\pi)^{-3} \int V^{1/2}(x)\, \chi(\vec{x}, k)\, dx \tag{VI.2}$$

and

$$\chi(\vec{x}, k) = |V(x)|^{1/2} + \int d^3y\, M(\vec{k}, \vec{y}; k)\, \chi(\vec{y}, k) \tag{VI.3}$$

where:

$$M(\vec{x}, \vec{y}; k) = -\frac{1}{4\pi} V_{\|}{}^{1/2}(x)\, \frac{e^{ik|x-y|-ik\hat{e}\cdot(x-y)}}{|x-y|} V^{1/2}(y) \tag{VI.4}$$

Remarks

1. We have suppressed the $\hat{e}$ dependence of χ and M.

2. The naturalness of $R \cap L^1$ is once again clear. If we had arrived at (VI.3) by formal manipulation of the Lippmann-Schwinger equation, we might demand $V \in L^1$ to assure the inhomogeneous term in (VI.3) is L^2 and $V \in R$ so that M is Hilbert-Schmidt.

3. We reiterate once more, that from *one point of view*, the prime physical result of this monograph is to justify from first principles the connection between T as defined by (V.22) and physical scattering and thereby the connection between the analyticity we discuss below and physical scattering.

We will analytically continue T by continuing M in the obvious way. We will then want to associate homogeneous solutions of $M\psi = \psi$ and solutions of $R_k\phi = \phi$, where R_k is given by (I.17), and thereby with bound states. To this end, we prove:

THEOREM VI.2. *Let* $K(x, y)$ *be a Hilbert-Schmidt kernel. Let* F *be a function which is measurable and never zero. Suppose* $F(x)K(x, y)F(y)^{-1}$ *is also Hilbert-Schmidt. Then any* L^2*-eigenfunction* ϕ *of* K $(K\phi = \phi)$ *has* $F\phi \in L^2$.

Proof. Let $B = \{\psi \in L^2 | F\psi \in L^2\}$. Norm B by $|||\psi||| = \|\psi\|_2 + \|F\psi\|_2$. B is complete by an elementary argument.[3] Let $C_m = \{x|\ |F(x)| < m\}$. Then $\cup C_m = R^3$ and C_m is an increasing family of sets, so we can find C_n with

$$\int_{R^3-C_n} dx \int dy[F(x)K(x,y)F(y)^{-1}]^2 = a_1 < 1 \qquad \text{(VI.5)}$$

$$\int_{R^3-C_n} dx \int dy|K(x,y)|^2 = a_2 < 1. \qquad \text{(VI.6)}$$

Let

$$K_1(x,y) = \chi_{C_n}(x)\, K(x,y)$$

$$K_2(x,y) = (1 - \chi_{C_n}(x))\, K(x,y).$$

If $\psi \in B$, $\|K_2\psi\|_2 \le a_2\|\psi\|_2$ and $\|FK_2\psi\| \le a_1\|F\psi\|_2$ (by VI.5) so $K_2 : B \to B$ and $|||K_2||| \le \max(a_2,a_1) < 1$. Since F is bounded on C_n, $\phi \in L^2$ implies $K_1\phi \in B$. Let (ϕ is the eigenfunction $K\phi = \phi$):

$$\eta = |||\ |||\text{-limit} \sum_{n=0}^{\infty} (K_2{}^n)(K_1\phi)$$

which exists and is in B. Then

$$(1-K_2)\eta = K_1\phi = (1-K_2)\phi \text{ (since } K\phi = \phi).$$

But (VI.6) implies $\|K_2\|_{L^2} < 1$ so $(1-K_2)^{-1}$ exists. Thus, $\phi = \eta \in B$. ∎

Remarks

1. This theorem holds on any measure space.
2. $K\phi = \phi$ implies $(FKF^{-1})(F\phi) = F\phi$. Thus this theorem (applied

---

3 If ψ_n is $|||\ |||$-Cauchy, then $\psi_n \to \psi$ in L^2 and $F\psi_n \to \phi$ in L^2. Since we can find subsequences convergent a.e., $\phi(x) = F(x)\psi(x)$ so $\psi \in B$ and $\psi_n \to \psi$ in $|||$

also with $K' = FKF^{-1}$, $K = GK' G^{-1}$, $G = F^{-1}$) implies that there is a one-one correspondence between homogeneous solutions of K and FKF^{-1}. We examine the way in which this "extends" Fredholm theory in the next Corollary.

3. Fredholm theory is so old and popular, that this theorem must appear somewhere in the literature although I know of no explicit reference.

4. In place of Theorem VI.2, Grossman and Wu [44, 45] use an interated integral equation method. Hunziker's approach [52] which involves cutting off the potential and taking limits is not unrelated to the method of proof we use.

5. One can find a quick alternate proof of Remark 2, by noticing F drops out of the Fredholm (or modified Fredholm) determinant for FKF^{-1}.

COROLLARY VI.3. *Let* K *and* F *be as in Theorem VI.2 and let* $N = FKF^{-1}$. *Then* $(1-N)^{-1}$ *exists if and only if* $(1-K)^{-1}$ *exists and*

$$(1-N)^{-1} = F(1-K)^{-1} F^{-1} \qquad \text{(VI.7)}^4$$

Proof. The simultaneous existence of the inverses is a consequence of the Fredholm alternative and Theorem VI.4 (Remark 2). (VI.7) is a direct consequence of the fact that F drops out of the Fredholm (or modified Fredholm determinantal solutions for the resolvent kernel. ∎

Remark

The net result of Theorem VI.2 and VI.3 is that a Fredholm integral resolvent canbe carried over without change to solve equations with the same kernel but with a non-L^2 inhomogeneous term. This is essentially what we are doing when we introduce χ in place of ψ.

---

4 We mean this in the sense that if $(1-N)^{-1}-1$ has kernel $\tilde{N}(x, y)$ and $(1-K)^{-1}-1$ has kernel $\tilde{K}(x, y)$, then $\tilde{N}(x, y) = F(x)\,\tilde{K}(x, y)\,F^{-1}(y)$.

THEOREM VI.4. *Let* V *be central and* $V \in L^1 \cap R$. *Let* $T_F(k)$ *be the forward* ($\theta = 0$) T *matrix defined with* $k > 0$ *and* $k^2 \notin \mathcal{E}$. *There exists a function* $\mathcal{T}_F(k)$ *meromorphic in the upper k-plane, so that:*

(a) *All the poles of* $\mathcal{T}_F(k)$ *are on the imaginary axis and are simple; there are at most finitely many of them.*

(b) *The poles occur at and only at points* k_0 *with* $k_0^2 = E_0 < 0$ *an eigenvalue of* H ($= H_0 + V$) *and the pole term is of the form* $R/k^2 - k_0^2$ *with*

$$R = (2\pi)^{-3}(\int V(x)\, \bar{\phi}_{|k|}(x) e^{+|k|\hat{e}\cdot x}\, dx)\, (\int V(y) \phi_{|k|}(y) e^{-|k|\hat{e}\cdot y}\, dy) \qquad \text{(VI.8)}$$

where $\phi_{|k|}$ *is the eigenvector of eigenvalue* E_0 *(if there are degenerate eigenvalues, we must sum over eigenfunctions in (VI.9)).*

(c) *For* k_0^2 *real and non-exceptional* $\lim_{k \to k_0} \mathcal{T}_F(k) = T_F(k_0)$ *and this limit is uniform on compact subsets of* $R - \mathcal{E}^{1/2}$.

(d) $\lim_{k \to \infty} \mathcal{T}_F(k) = T_{Born} \equiv (2\pi)^{-3} \int V(x)\, d^3x$. *This limit includes* k *real if* $\mathcal{T}_F(k)$ *is defined to be its boundary value* $T_F(k)$ *on* $R - \mathcal{E}^{1/2}$. *(Since* $\mathcal{E}$ *is compact,* k *real and large is in* $R - \mathcal{E}^{1/2}$*).*

(e) $\mathcal{T}_F(-\bar{k}) = \overline{\mathcal{T}_F(k)}$

(i.e., symmetry about the imaginary axis).

Proof. Define the kernel $M(\vec{x}, \vec{y}; k)$ by (VI.4), for $\operatorname{Im} k \geq 0$. Since $|x-y| - \hat{e}\cdot(x-y) \geq 0$ for all x, y, we have M Hilbert-Schmidt and analytic for all k in the upper half-plane. Thus, by the analytic Fredholm theorem (Theorem A.27), $(1 - M_k)^{-1}$ exists[5] as a meromorphic function. Define:

$$\mathcal{T}_F(k) = (2\pi)^{-3} \langle V^{1/2}, (1 - M_k)^{-1} V_{\|}^{1/2} \rangle.$$

---

[5] We will see that $\|M_k\| \to 0$ as $k \to \infty$, so we can eliminate the possibility that $(1 - M_k)$ is nowhere invertable.

The meromorphy of $\mathcal{T}_F$ is evident. For k imaginary, M is real and thus, so is $\mathcal{T}_F$. By analytic continuation, (e) follows. By the Klein-Zemach theorem (Theorem I.23) $\|M_k\| \to 0$ as $k \to +\infty$ along the real axis and similar arguments show that $\|M_k\| \to 0$ as $k \to \infty$ in any manner. Thus (d) is true. By definition of $\mathcal{E}$ (and Theorem VI.2)

$$(1 - M_k)^{-1} V_{\|}^{1/2} \overset{L^2}{\to} \chi(\cdot, k_0)$$

as $k \to k_o \notin \mathcal{E}^{1/2}$ and the convergence is uniform on compacts so (c) follows. The Fredholm alternative tells us that $(1 - M_k)^{-1}$ only fails to exist when $M_k\phi = \phi$ has a solution. We have

$$M(x, y; k) = e^{-ik\hat{e}\cdot x} R_k(x, y)\, e^{+ik\hat{e}\cdot y}$$

so this can only happen if $R_k(x, y)$ has a solution (by Theorem VI.2) and thus only if k^2 is an eigenvalue (by Theorem III.4). To complete the proof, we need only show that the poles are simple and have "residues" given by (VI.8).

By Corollary VI.3:

$$(1 - M_k)^{-1} = e^{-ik\hat{e}\cdot x} (1 - R_k)^{-1} e^{+ik\hat{e}\cdot y} \tag{VI.9}$$

for k away from the pole. But, by arguments similar to those in Section II.9:

$$(1 - R_k)^{-1} = (1 - V_{\|}^{1/2} (E - H_o)^{-1} V^{1/2})^{-1} = 1 + V_{\|}^{1/2} (E - H)^{-1} V^{1/2}.$$

As a result, $(1 - R_k)^{-1}$ has a pole of the form $V_{\|}^{1/2} PV^{1/2}/E{-}E_0$. Thus the pole is simple and the residue is given by

$$R = (2\pi)^{-3} \langle V^{1/2}, e^{+|k|\hat{e}\cdot x} V_{\|}^{1/2} PV^{1/2} e^{-|k|\hat{e}\cdot y} V_{\|}^{1/2} \rangle$$

which is (VI.8). ∎

COROLLARY VI.5. *Let* $V \in R \cap L^1$ *and central. Let* $f_{Phys}(E, \theta = 0)$ *be the forward scattering amplitude. Then there is a function* $f(E, \theta = 0)$ *meromorphic in the canonically cut plane so that:*

(a) f *has poles only on the negative axis. These poles are simple and occur only at bound state energies.*

(b) *If* E *is not exceptional*

$$f_{Phys.}(E, \theta = 0) = \lim_{\varepsilon \downarrow 0} f(E + i\varepsilon, \theta = 0)$$

in which the limit uniform on compact subsets of $R - \mathcal{E}$.

(c) $\lim_{E \to \infty} f(E, \theta = 0) = -(4\pi)^{-1} \int V(x)\, d^3x \equiv f_{Born}$

(d) $f(\bar{E}, \theta) = \overline{f(E, \theta)}$

(e) *There exists a distribution* "Imf(E)" *which for* E *non-exceptional is the continuous function* $\operatorname{Im} f(E)$ *so that*

$$\operatorname{Re} f(E) = \frac{\mathcal{P}}{\pi} \int_0^\infty \frac{\text{``Im''} f(E')}{E' - E}\, dE' + f_{Born} + \sum_{j=1}^{n} \frac{r_j}{E_j - E}$$

whenever E *is non-exceptional.* ∎

Appendix to VI.2. Behavior of Eigenfunctions at Infinity

Theorem VI.2 is tailor-made to yield information about the behavior of negative energy eigenfunctions near $x = \infty$. Since there are complications in the Rollnik case, we first consider L^2-potentials where the idea is quite clear.

THEOREM VI.6. *Let* $V \in L^2$. *Let* $(-\Delta + V)\Psi = E\Psi$ *with* $E = -\kappa_0^2 < 0$. *Let* $\kappa < \kappa_0$. *Then* $e^{\kappa|x|}\Psi(x) \in L^2$ [*so* Ψ *is more-or-less* $O(e^{-\kappa_0|x|})$].

Proof. Since $V \in L^2$, $\Psi = (E - H_0)^{-1} V\Psi$ so Ψ is a solution of

$$\Psi = \int -\frac{e^{-\kappa_0|x-y|}}{4\pi|x-y|} V(y)\, \Psi(y)\, d^3y$$

which has a Hilbert-Schmidt kernel. Let

$$M_\kappa(x, y) = -\frac{1}{4\pi}\frac{e^{-\kappa_0|x-y|}}{|x-y|} e^{+\kappa|x|} e^{-\kappa|y|} V(y)$$

Since

$$|x-y| + |y| - |x| \geq 0,$$

$$|M_\kappa(x, y)| \leq (4\pi)^{-1} |x-y|^{-1} e^{-(\kappa_0-\kappa)|x-y|} V(y)$$

Thus M_κ is Hilbert-Schmidt so this theorem is implied by Theorem VI.2. ∎

The same method clearly shows $V^{1/2}(x)\,\Psi(x)\,e^{+\kappa|x|} \in L^2$ when $V \in R$. Since V may vanish at large distances, this is not terribly satisfactory. However a technical trick allows us to extend VI.6 to the Rollnik case:

THEOREM VI.7.* *Let* $V \in R$ *and let* $H\Psi = E\Psi$ *with* $E = -\kappa_0^{\,2} < 0$. *Let* $\kappa < \kappa_0$. *Then* $e^{\kappa|x|}\Psi(\vec{x}) \in L^2$.

Proof. We have $V_{\|}^{1/2}\Psi = [V_{\|}^{1/2}(E - H_0)^{-1}\, V^{1/2}]\, V_{\|}^{1/2}\Psi$ and for almost all x:

$$\Psi(x) = -\frac{1}{4\pi}\int \frac{e^{-\kappa_0|x-y|}}{|x-y|} V(y)\,\Psi(y)\,d^3y\,.$$

Let $A(x) = \sqrt{|V(x)| + e^{-2\varepsilon|x|}}$. Then $A^2 \in R$ and $|V/A(y)| \leq |V(y)|^{1/2}$, so $|V|^2/A^2 \in R$. Moreover $A\Psi = [A(E - H_0)^{-1}\,(V/A)](A\Psi)$. This last kernel is Hilbert-Schmidt and $A\Psi \in L^2$ ($A^2 \in R$ and $\Psi \in D(H_0^{\,1/2})$ imply $A\Psi \in L^2$). Since $e^{+\kappa|x|}[A(E - H_0)^{-1}\, V/A]\, e^{-\kappa|y|}$ is also Hilbert-Schmidt for any $\kappa < \kappa_0$. we see that $e^{+\kappa|x|}A\Psi \in L^2$ so $e^{(\kappa-\varepsilon)|x|}\Psi \in L^2$ by Theorem VI.2. Since $\varepsilon > 0$ and $\kappa < \kappa_0$ are arbitrary, the theorem is proven. ∎

VI.3. *Analyticity of the Non-Forward Amplitude*

We study analyticity of the non-forward amplitude for potentials of finite range. The variables and general results are exactly those of Hunziker [52] although the conditions on V are different; Grossman-Wu

---

* A. O'Connor (Princeton Thesis, in preparation) has extended this theorem to $V \in R + (L^\infty)_\varepsilon$ and to prove exponential falloff of wave functions in the n-body case whose energy is below the continuum limit.

also have similar results. Finally, we follow Tiktopoulos' unpublished treatment nearly exactly (modulo technical details). The amplitude $f(\vec{\kappa}, \vec{\kappa}')$ is really only a function of $E = k^2$ $(= k'^2)$ and $\cos\theta = \hat{k}\cdot\hat{k}'$. We introduce the alternate variables P and Δ defined by (Hunziker):[6]

$$\underset{\sim}{P} = \frac{1}{2}(k' + k) = P\hat{e}$$

$$\underset{\sim}{\Delta} = \frac{1}{2}(k' - k) = \Delta\hat{e}' \qquad \text{(VI.9)}$$

so $e\cdot e' = 0$, and $E^2 = P^2 + \Delta^2$; $\cos\theta = P^2 - \Delta^2/P^2 + \Delta^2$. Because of the rotational invariance of f and the condition $e\cdot e' = 0$, independent of P and Δ, we can fix e and e' once and for all. We will write $f(k, \Delta)$ for the obvious function. Let V have inverse range C and let $\alpha < C$;[7] $-1 \leq \lambda \leq 1$. Following Hunziker, define:[8]

$$\chi_{\lambda,\alpha}(\vec{x}, k) = \exp(+i\lambda k(\hat{e}\cdot x) - \frac{1}{2}\alpha|x|)\,\Psi(\vec{x}, \vec{k}) \qquad \text{(VI.10)}$$

where:

$$\vec{k} = P\hat{e} - \Delta\hat{e}' \qquad \text{(VI.11)}$$

and Ψ is the familiar solution of (IV.17). Then χ solves the integral equation (when $k \notin \mathcal{E}$):

$$\chi(\vec{x}, k) = \chi_0(\vec{x}, k) + \int M_{\lambda,\alpha}(\vec{x}, \vec{y}; k)\,\chi(\vec{y}, k)\,d^3y \qquad \text{(VI.12)}$$

where:

$$\chi_0 = \exp[-i\Delta\hat{e}'\cdot x + i(P + \lambda k)(\hat{e}\cdot x) - \frac{1}{2}\alpha|x|]|V(x)|^{1/2} \qquad \text{(VI.13)}$$

---

6 E, Δ and P play the role of the familiar Mandelstam variables s, t and u. In fact, in the C.M. system for equal mass scatting, $s = 4(\mu^2 + E^2)$; $t = -4\Delta^2$ and $u = -4P^2$ in terms of our Δ and P.

7 Thus $\int dx\; e^{\alpha|x|}\,|V(x)| < \infty$ and $\int dx\,dy\,|x-y|^{-2}|V(x)|e^{\alpha|x|}\,|V(y)|e^{\alpha|y|} < \infty$.

8 We suppress the dependence on P and eventually on λ and α.

$$M = -[4\pi|x-y|]^{-1} e^{-\frac{1}{2}\alpha|x|} V_{\|}^{\frac{1}{2}}(x) e^{ik|x-y|+i\lambda k e\cdot(x-y)} V^{\frac{1}{2}}(y) e^{\frac{1}{2}\alpha|y|} . \quad \text{(VI.14)}$$

Moreover:

$$\begin{aligned} f(k, \Delta) &= -(4\pi)^{-1} \int d^3x\, e^{-i\vec{k}'\cdot\vec{x}} V^{\frac{1}{2}}(x)\, \Psi(\vec{x}, \vec{k}) \\ &= -(4\pi)^{-1} \int d^3x\, e^{-i\Delta\hat{e}'\cdot x - i(P+\lambda k)\hat{e}\cdot x + \frac{1}{2}\alpha|x|} V^{\frac{1}{2}}(x)\, \chi(\underset{\sim}{x}, k) \end{aligned}$$

so that:

$$f(k, \Delta) = f_{Born}(\Delta) - (4\pi)^{-1} \int d^3x\, e^{-i\Delta e'\cdot x - i(P+\lambda k)e\cdot x - \alpha|x|}$$

$$(V^{\frac{1}{2}}(x)\, e^{\frac{1}{2}\alpha|x|})\, [(\chi - \chi_0)\, e^{\alpha|x|}] \quad \text{(VI.15)}$$

where

$$f_{Born}(\Delta) = -(4\pi)^{-1} \int e^{-2i\Delta\hat{e}'\cdot x}\, V(x)\, dx . \quad \text{(VI.16)}$$

The kernel M is clearly Hilbert-Schmidt in the upper-half plane and in fact $\int |e^{\alpha|x|} M(x, y)|^2 < \infty$ so that $(\chi - \chi_0)\, e^{\alpha|x|} \in L^2$ when $\chi_0 \in L^2$. The inhomogeneous term is bounded by

$$|V(x)|^{\frac{1}{2}} \exp[\{-\tfrac{\alpha}{2} + \sqrt{(\mathrm{Im}\Delta)^2 + |\mathrm{Im}(P + \lambda k)|^2}\}\, |x|]$$

so we have an L^2-inhomogenous term if

$$\sqrt{(\mathrm{Im}\Delta)^2 + |\mathrm{Im}(P + \lambda k)|^2} \leq \alpha . \quad \text{(VI.17)}$$

In this case the exponential in (VI.15) is also bounded (since $\hat{e}\cdot\hat{e}' = 0$) and so $f(k, \Delta) - f_{Born}(\Delta)$ can be extended to the region given by (VI.17); i.e., analogously to Theorem VI.4, we can prove:

THEOREM VI.8. *Let* A *be the interior of the region of complex* k *and* Δ *obeying* (VI.17) *for some* $-1 \leq \lambda \leq 1$. *Then there is a function* $\tilde{f}(k, \Delta)$ *in this region so that:*

(a) *All the poles of* f *are simple poles occurring either at values of* k *with* $k^2 \leq 0$ *an eigenvalue or at exceptional values of* $k^2 \geq 0$.[9]

(b) *If* (k_o, Δ_o) *is a non-expectional physical value,*[10] *then*

$$\tilde{f}(k_o, \Delta_o) = f_{Phys}(k_o, \Delta_o) - f_{Born}(\Delta_o).$$

(c) *If* $(k, \Delta) \in A$ *for fixed* Δ *and a set of* k's $\to \infty$, *then*

$$\lim_{k \to \infty} \tilde{f}(k, \Delta) = 0$$

(d) $f(-\bar{k}, \Delta) = \overline{f(k, \Delta)}$ *for* Δ *real, and suitable to put* $k, \Delta \in A$. ∎

Remarks.

1. The occurrence of $P = \sqrt{k^2 - \Delta^2}$ could a priori produce sigularities at $k^2 = \Delta^2$. However, $P \to -P$ is equivalent to $\hat{e} \to -\hat{e}$ which doesn't change f. Similarly $\Delta \to -\Delta$ doesn't change f.

2. f_{Born} is analytic in a different region than A, viz. $\{\Delta \mid |\text{Im}\,\Delta| < \frac{1}{2}a\}$. Thus in the overlap of this region and A, we can make statements about the full analytically-continued amplitude; e.g., $\lim_{k \to \infty} f(k, \Delta) = f_B(\Delta)$.

Let us rewrite the crucial condition (VI.17).

LEMMA VI.9. (VI.17) *holds for some* $-1 \leq \lambda \leq 1$ *if and only if*

$$|\text{Im}\,\Delta| \leq a \qquad \text{(VI.18a)}$$

$$|\text{Im}\,\sqrt{k^2 - \Delta^2}| - |\text{Im}\,k| \leq \sqrt{a^2 - (\text{Im}\,\Delta)^2}. \qquad \text{(VI.18b)}$$

Proof. Case 1. $|\text{Im}\,\sqrt{k^2 - \Delta^2}| \leq |\text{Im}\,k|$.

Then (VI.18b) holds automatically, so we must show (VI.18a) is equivalent to (VI.17) in this case. Let (VI.17) hold. Then (VI.18a) holds. Conversely, let (VI.18a) hold and let

9 Since V has finite inverse range, $\mathcal{E}$ is a finite discrete set.

10 That is $k_o^2 \notin \mathcal{E}$; $0 \leq \Delta \leq k_o^2$.

$\lambda = -\text{Im}\sqrt{k^2 - \Delta^2}/|\text{Im } k|$ so $|\lambda| < 1$. Then $\text{Im}(P + \lambda k) = \text{Im}(\sqrt{k^2 - \Delta^2} + \lambda k) = 0$ so (VI.17) holds.

Case 2. $|\text{Im}\sqrt{k^2 - \Delta^2}| \geq \text{Im } k$.

Let (VI.17) hold. (VI.18a) follows trivially and

$$|\text{Im}\sqrt{k^2 - \Delta^2}| - |\text{Im } k| \leq |\text{Im}\sqrt{k^2 - \Delta^2}| - |\lambda|\,|\text{Im } k|$$

$$\leq |\text{Im}(P + \lambda k)| \leq \sqrt{a^2 - (\text{Im }\Delta)^2}$$

so (VI.18b) holds. Conversely, let (VI.18a, b) hold; let $\lambda = -\text{sgn}(\text{Im}\sqrt{k^2 - \Delta^2})$. Then $|\text{Im }\Delta|^2 + |\text{Im }(\sqrt{k^2 - \Delta^2} + \lambda k)|^2 = (\text{Im }\Delta)^2 + [(\text{Im}\sqrt{k^2 - \Delta^2}) - (\text{Im } k)]^2 \leq a^2$ so (VI.17) holds. ∎

THEOREM VI.10 (Fixed Δ analyticity). (VI.18a, b) *holds for all* k *(with* $\text{Im } k > 0$ *which we are always supposing) if and only if* $|\Delta| \leq a$.

Proof. Let $f(k) = |\text{Im}\sqrt{k^2 - \Delta^2}| - \text{Im } k$ be defined in the upper half plane. As $k \to \infty$, a simple computation shows $f(k) \to 0$ so f(k) takes its maximum value at a finite point. Since $\pm\,\text{Im}\sqrt{k^2 - \Delta^2} - \text{Im } k$ are harmonic functions, this *maximum* value cannot be at an interior point, so it must be on the real axis, where $f(k) = |\text{Im}\sqrt{k^2 - \Delta^2}|$. This clearly has its maximum value at $k = 0$, where $f(k) = |\text{Re }\Delta|$. Thus (VI.18b) holding for all k is equivalent to $(\text{Re }\Delta)^2 + (\text{Im }\Delta)^2 \leq a^2$ and in that case (VI.18a) is subsidiary to (VI.18b). ∎

COROLLARY VI.11 (Fixed Δ-dispersion relations). *Let* $V \in R \cap L^1$ *have finite inverse range* C. *Let* $|\Delta| < C$. *Then there exists a function* $g(E, \Delta)$ *meromorphic in the canonically cut plane so that:*

(a) f *has poles only on the negative axis. These poles are all simple and occur only at bound state energies.*

(b) *If* $\sqrt{E}$ *is non-exceptional,* $\lim_{\varepsilon \downarrow 0} g(E + i\varepsilon, \Delta)$ *exists. If moreover,* Δ *is physical* (*i.e.*, Δ real *and* $|\Delta| \leq E$), *then* $g(E + i\varepsilon, \Delta) + f_{\text{Born}}(\Delta)$ *is the physical amplitude.*

(c) $\lim_{\substack{E \to \infty \\ \Delta \text{ fixed}}} g(E, \Delta) = 0.$

(d) $g(\overline{E}, \Delta) = \overline{g(E, \Delta)}$ for Δ real

$g(E, -\Delta) = g(E, \Delta).$

(e) *There is a distribution* "Im f(E, Δ)" *which for* E *non-exceptional is the continuous function* Im f(E) *so that:*

$$\operatorname{Re} f(E, \Delta) = \frac{\mathcal{P}}{\pi} \int_0^\infty \frac{\text{“}\operatorname{Im} f(E, +\Delta)\text{”}}{E - E'} + f_{\text{Born}}(\Delta) - \sum_{j=1}^{n} \frac{r_j(\Delta)}{E_j - E}$$

whenever E *is non-exceptional and* Δ *is real.* ∎

Remarks.

1. The physical region begins at $E \geq |\Delta|$ so the dispersion integral is *not* completely in the physical region (unlike the forward case).

2. We will see in the next section that the possible singularities in $\mathcal{E}$ are removable.

THEOREM VI.12 (Fixed physical k-analyticity). *Let* k *be fixed, real and positive. Then (VI.17) holds if and only if* $z = \cos\theta$ *is in the ellipse with foci* ± 1 *and semi-major axis* $1 + 2k^{-2}\alpha^2$ (Lehmann-ellipse):

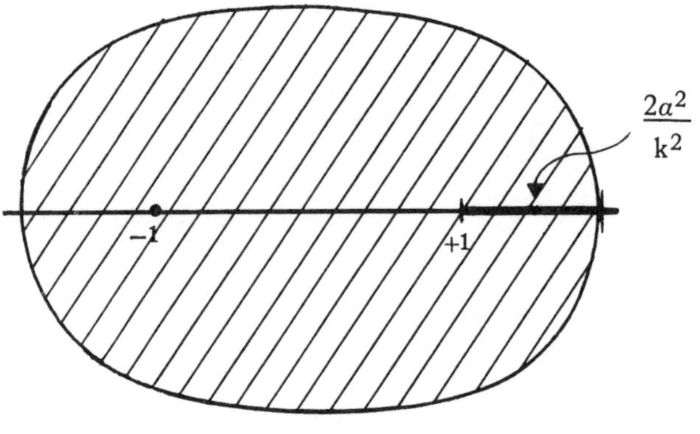

Proof. Since $\operatorname{Im} k = 0$, (VI.17) is equivalent to $(\operatorname{Im}\Delta)^2 + (\operatorname{Im}\sqrt{k^2 - \Delta^2})^2 \leq \alpha^2$ since $\Delta^2 = \frac{k^2}{2}(1-\cos\theta)$, we see that this is equivalent to

$$(\mathrm{Im}\ k\sqrt{\tfrac{1-z}{2}})^2 + (\mathrm{Im}\ k\sqrt{\tfrac{1+z}{2}})^2 \leq a^2$$

$$(\mathrm{Im}\sqrt{1-z})^2 + (\mathrm{Im}\sqrt{1+z})^2 \leq \frac{2a^2}{k^2} \qquad \text{(VI.19)}$$

Since $(\mathrm{Re}\ w)^2 - (\mathrm{Im}\ w)^2 = \mathrm{Re}(w^2)$ we have $|w|^2 = \mathrm{Re}(w^2) + 2(\mathrm{Im}\ w)^2$, so that (VI.19) is equivalent to:

$$|1-z| + |1+z| \leq \frac{4a^2}{k^2} + \mathrm{Re}(1-z) + \mathrm{Re}(1+z)$$

i.e.,

$$|1-z| + |1+z| \leq 2(1 + 2a^2/k^2)$$

which is the stated ellipse. ∎

Even when k is not physical, we can say something:

THEOREM VI.13 (Fixed non-physical k-analyticity). *Let* k *be fixed,* $0 < \mathrm{Im}\ k < a$. *Then* (VI.17) *holds for* $z = \cos\theta$ *in the ellipse with foci* ± 1 *and semi-major axis* $1 + \dfrac{2[a^2 - (\mathrm{Im}\ k)^2]}{k^2}$.

Remark

This condition is sufficient but not necessary for (VI.17) to hold.

Proof. A sufficient condition for (VI.17) is

$$(\lambda = 0),\ \mathrm{Im}\sqrt{k^2 + \Delta^2} + (\mathrm{Im}\ \Delta)^2 < a^2$$

or equivalently,

$$(\mathrm{Im}\ k\sqrt{1-z})^2 + (\mathrm{Im}\ k\sqrt{1+z})^2 < 2a^2 \qquad \text{(VI.20)}$$

Using $|w|^2 = \mathrm{Re}(w^2) + 2(\mathrm{Im}\ w)^2$ as in Theorem VI.12, we see that (VI.20) holds if

$$|k|^2(|1-z| + |1+z|) < 4a^2 + 2\mathrm{Re}\ k = 4a^2 + 2|k|^2 - 4|\mathrm{Im}\ k|^2$$

or

$$|1-z| + |1+z| < 2[1+2|k|^{-2} (a^2-|\mathrm{Im}\, k|^2)].$$

VI.4. *The Partial Wave Expansion*

As a final element of our presentation of analyticity, we exhibit those properties of the partial wave expansion which one can abstract from our discussion of the full amplitude. In the L^2-case, one can obtain strong results by a direct study of the radial Schrödinger equation (see, e.g., De Alfaro and Regge [T1]). We will not examine in detail to what extent this approach carries over to the Rollnik case although we see no reason why it shouldn't.

The basic fact needed to derive properties of the partial wave amplitude from the full amplitude is:

THEOREM VI.14. *Let* f(z) *be a function analytic in an ellipse with* foci ±1. *Then* f(z) *has a Legendre expansion:*

$$f(z) = \sum_{\ell=0}^{\infty} (2\ell+1)\, a_\ell P_\ell(z) \tag{VI.21}$$

convergent uniformly on compact subsets of the open ellipse. (VI.21) *is pointwise convergent in the interior of the largest ellipse of analyticity and strictly divergent outside the ellipse. The coefficients* a_ℓ *are given by*

$$a_\ell = \frac{1}{2} \int_{-1}^{1} dz\, f_\ell(z)\, P_\ell(z) \tag{VI.22}$$

$$= \frac{1}{2\pi i} \int_C f(z)\, Q_\ell(z)\, dz \tag{VI.23}$$

where Q_ℓ *is the associated Legendre function and* C *a contour inside the ellipse cut by* (−1,1) *and winding once about the cut. The* a_ℓ *obey:*

$$\overline{\lim}\ |a_\ell|^{1/\ell} \leq A + \sqrt{A^2-1} \tag{VI.24}$$

where A *is the semi-major axis of the ellipse of analyticity. Moreover,*

(VI.24) *is a sufficient condition for* (VI.21) *to converge in the ellipse of semi-major axis* A *(and foci* ± 1*).*

Proof. We present a sketch of the proof of crucial elements of this result in an appendix to this section. ∎

COROLLARY VI.15. *Let* V *be a potential in* $R \cap L^1$ *of finite range. Then for any physical* z $(-1 \le z \le 1)$ *and any non-exceptional value of* k, *the physical amplitude* f(k, z) *has a uniformly convergent partial wave expansion* $f(k, z) = \sum_{\ell=0}^{\infty} (2\ell+1) f_\ell(k) P_\ell(z)$. *The functions* $f_\ell(k) - f_\ell^{(\mathrm{Born})}(k)$ *are the boundary values of functions analytic in the region* $0 < \mathrm{Im}\, k < a$. *(The* $f_\ell^{(\mathrm{Born})}(k)$ *are analytic in* $0 < \mathrm{Im}\, k < \frac{1}{2} a$*). Moreover:*

(a) $\overline{f_\ell(k)} = f_\ell(-\bar{k})$

(b) *For* k *real and non-exceptional*

$$\overline{\lim}\, |f_\ell(k)|^{1/\ell} \le A(k) + \sqrt{A(k)^2 - 1} \quad \textit{with} \quad A(k) = 1 + \frac{2a^2}{2k^2}$$

(c) *For* $\mathrm{Im}\, k > 0$, $\overline{\lim}\, |f_\ell(k)|^{1/\ell} \le A(k)$ *with*

$$A(k) = 1 + 2|k|^{-2} [a^2 - 4(\mathrm{Im}\, k)^2].$$

Proof. From Theorems VI.8, VI.12, and VI.13, $f(k, z) - f_{\mathrm{Born}}(\Delta)$ has sufficient properties to yield the stated properties of $f_\ell(k) - f_\ell^{(\mathrm{Born})}(k)$ via Theorem VI.16. The Born term is given by:

$$f_{\mathrm{Born}}(k, z) = -\frac{1}{4\pi} \int V(x) \exp\left[-2i(e' \cdot x) k \sqrt{(1-z)/2}\right]$$

and is thus analytic if $|\mathrm{Im}\, k \sqrt{1-z}|^2 < \frac{1}{2} a^2$. By an analysis identical to that following (VI.20), this region contains an ellipse with foci ± 1 and semi-major axis $1 + \frac{1}{2} |k|^{-2}(a^2 - 4|\mathrm{Im}\, k|^2)$. ∎

We have been promising to show that when V is of finite range, the exceptional points "decouple" from the scattering amplitude. Not surprizingly, the proof of this fact depends on partial wave unitarity:

LEMMA VI.16. *Suppose* $V \in R$ *has finite range. Let* k *be real and non-exceptional. Then* $\operatorname{Im} f_\ell(k) = k|f_\ell(k)|^2$. *Equivalently,* $f_\ell(k) = \frac{1}{k}(e^{i\delta_\ell}\sin\delta_\ell)$, *with* $\delta_\ell(k)$ *real.*

Proof. Since the details of the computations are standard, we only sketch the proof. From Theorem V.10 and the unitarity of S, we immediately obtain:

$$\operatorname{Im} T(k,k') = \pi \int d^3k''\, \delta(k''^2 - k^2)\, T(k,k'')\, T(k'',k') \qquad \text{(VI.25)}$$

for almost all (k, k') with $k^2 = (k')^2$. Using the uniform convergence of the partial wave expansions and the addition formula for Legendre polynomials, we obtain $\operatorname{Im} f_\ell = k|f_\ell|^2$ a.e., in k. By analytic continuation,[11, 12] the equality extends to all real non-exceptional k.[13] ∎

As two immediate consequences of unitarity, we have:

THEOREM VI.17. $f_\ell(k)$ *has a meromorphic continuation to the entire strip* $|\operatorname{Im} k| < \frac{1}{2} a$ *when* V *has finite range* a^{-1}.

Proof. Since unitarity has the form $f_\ell(-k) = f_\ell(k)\,[1 + 2ik\, f_\ell(k)]^{-1}$, this follows from Theorem VI.15.

THEOREM VI.18. *Let* $V \in R$ *have finite range. Then any exceptional points* $(k \neq 0)$ *do not couple to the on-shell scattering amplitude.*

---

[11] We will shortly see that $f_\ell(k)$ is analytic in a neighborhood of $R - \mathcal{E}$ so continuation from the real axis presents no technical difficulties.

[12] Despite its appearance $\operatorname{Im} f = k|f|^2$ is an analytic statement, for $f_\ell(k) = f_\ell(-k)$ when k is real. Thus unitarity has the form $f_\ell(k) - f_\ell(-k) = (2ik) f_\ell(-k) f_\ell(k)$.

[13] We will see below that it extends to all k.

Explicitly:

(a) *The partial wave amplitudes have removable singularities at the exceptional points.*

(b) *The scattering amplitude* $f(k, \Delta)$ *has a removable singularity at the exceptional points, when* z *is in the Lehmann ellipse.*

Proof. By shrinking the Lehmann ellipse a little, we can "save-up" some of the exponential fall-off of V, to allow $\mathrm{Im}\, k$ to have a small imaginary part. We can thereby obtain $(1-R_k)^{-1}$ and $(1-M_k)^{-1}$ as *meromorphic* functions in a neighborhood of any $k_o \in \mathcal{E}$.[14] Thus

$$f(k, \Delta) = \frac{R_n(\Delta)}{(k-k_o)^n} + \ldots + \frac{R_1(\Delta)}{k-k_o} + A(k, \Delta)$$

where the $R_i(\Delta)$ are analytic for Δ in the Lehmann ellipse and $A(k, \Delta)$ is analytic at k_o. Partial wave projecting,

$$f_\ell(k) = \frac{R_{n,\ell}}{(k-k_o)^n} + \ldots + \frac{R_{1,\ell}}{(k-k_o)} + A_\ell(k).$$

Since $|f_\ell(k)| \leq k^{-1}$ for k near k_o,[15] we must have $R_{n,\ell} = \ldots = R_{1,\ell} = 0$, for all ℓ. Thus $R_n(\Delta) = \ldots = R_1(\Delta) = 0$ so that (a) and (b) are proven. ∎

Appendix to VI.4. Convergence of Legendre Series

For the reader's convenience, we present a discussion of the convergence of Legendre series. Our proof is simpler (proceeding from first principles) than the one found in Whitaker and Watson [T27], pp. 318-323. We begin with a "geometric" lemma which is the source for all the ellipses!:

---

14 We already know by this sort of argument that $\mathcal{E}$ is discrete.

15 Unitarity implies $|1 + 2ik\, f_\ell(k)| = 1$.

LEMMA VI.19. *The curves* $|\mathrm{Im}\ \theta| = \kappa$ *in the* $z = \cos\theta$ *plane are ellipses with foci at* ± 1 *and* $\cosh\kappa$ *as semi-major axis.*

Proof. Let $\theta = x + iy$. Then:

$$\exp(\pm i\theta) = \exp(\mp y)\exp(\pm ix)$$

so

$$\cos\theta = (\cos x)(\cosh y) + i(\sin x)(\sinh y).$$

Thus $|y| = \kappa$ is equivalent to:

$$\frac{[\mathrm{Re}(\cos\theta)]^2}{\cosh^2\kappa} + \frac{[\mathrm{Im}(\cos\theta)]^2}{\sinh^2\kappa} = 1. \blacksquare$$

LEMMA VI.20. $\overline{\lim\limits_{\ell\to\infty}}\ |P_\ell(z)|^{1/\ell} = [\min|e^{\pm i\theta}|]^{-1} = e^{+|\mathrm{Im}\theta|}$

where $z = \cos\theta$. *For any* $H > 0$, $P_\ell(z)H^\ell$ *is uniformly bounded on any z-compact inside the ellipse* $|\mathrm{Im}\ \theta| = \ln H$.

Proof. The $P_\ell(z)$ are defined by the generating function:

$$\sum_{\ell=0}^{\infty} h^\ell P_\ell(z) = (1-2hz+h^2)^{-\frac{1}{2}}. \tag{VI.26}$$

The function on the left hand side has singularities at

$$h = z \pm \sqrt{z^2-1} = \cos\theta \pm i\sin\theta = e^{\pm i\theta}.$$

Thus the series has $\min|e^{\pm i\theta}|$ as a radius of convergence. But, by Hadamard's theorem, the radius of convergence is $[\overline{\lim}\ |P_\ell(z)|^{1/\ell}]^{-1}$. This proves the first part of the lemma. If C is a compact inside the ellipse, $H < e^{-|\mathrm{Im}\ \theta(z)|}$ for all $z \in C$, i.e., H lies inside the radius of convergence of (VI.26). Thus $\sup\limits_{z\in C;\ |h|=H} |(1-2hz+h^2)^{-\frac{1}{2}}| = k < \infty$. By Cauchy's estimate,

$$|P_\ell(z)| = \frac{1}{2\pi}| \int_{|h|=H} h^{-\ell-1} (1-2hz+h^2)^{-\frac{1}{2}}\, dh \le (\frac{k}{2\pi H})\, \frac{2\pi}{|H|^\ell}.$$

LEMMA VI.21. *Let* $Q_\ell(z) = \frac{1}{2}\int_{-1}^{1} \frac{P_\ell(z')\,dz'}{z-z'}\,[z \notin (-1, 1)]$.

Then $\overline{\lim}\, |Q_\ell(z)|^{1/\ell} = [\max_{\pm} |e^{\pm i\theta}|]^{-1} = e^{-|\mathrm{Im}\,\theta|}$ *where* $z = \cos\theta$. *For any* $H > 0$, $Q_\ell(z)H^\ell$ *is uniformly bounded on any z-compact outside the ellipse* $|\mathrm{Im}\,\theta| = \ln H$.

Proof. For compact subsets of $\{h| \; |h| < 1\}$, the convergence of (VI.26) is uniform in h and $z \in (-1, 1)$. Thus we have

$$\sum_{\ell=0}^{\infty} h^\ell Q_\ell(y) = \frac{1}{2}\int_{-1}^{1} \frac{dz}{y-z}(1 - 2hz + h^2)^{-\frac{1}{2}} \equiv G(h, y)$$

where

$$G(h, y) = \frac{1}{u}\log\left[\frac{y-h+u}{(y^2-1)^{\frac{1}{2}}}\right] \tag{VI.27}$$

[16]

with

$$u = (1 - 2hy + h^2)^{\frac{1}{2}}.$$

---

16 For

$$G(h, y) = \frac{1}{2}\int_{1-h}^{1+h} dv[hy-hz]^{-1}\ [v = (1-2hz+h^2)^{\frac{1}{2}}]$$

$$= \int_{1-h}^{1+h} \frac{dv}{v^2-u^2} = \frac{1}{2u}\Big|_{z=1}^{z=-1} \ln(\frac{v-u}{v+u})$$

$$= \frac{1}{2u}\ln\left[\frac{(y+1)}{(y-1)}\,\frac{(u+1-h)^2}{(u+1+h)^2}\right]$$

$$= \frac{1}{2u}\ln\left[\frac{(y+1)}{(y-1)}\,\frac{[(u-h)^2-1]^2}{u^2-(1+h)^2}\right]$$

$$= \frac{1}{2u}\ln\left[\frac{(y+1)}{(y-1)}\,\frac{(2h^2-2hy-2uh)^2}{[2h(y+1)^2]}\right] = \frac{1}{u}\ln\left[\frac{y-h+u}{(y^2-1)^{\frac{1}{2}}}\right]$$

As in the proof of Lemma VI.20, we immediately see that

$$\overline{\lim}\ |Q_\ell(y)|^{1/\ell} = [R(y)]^{-1}$$

where $R(y)$ is the radius of convergence of $G(h, y)$ for y fixed. We have:[17]

$$G(h, y) = \frac{1}{2u} \ln \left[\frac{1-(h-u)^2}{1-(h+u)^2} \right]. \tag{VI.28}$$

Consider first the argument of the logarithm. It is 0 or ∞ only if $u = \pm h \pm 1$, i.e., only if $u^2 = h^2 \pm 2h + 1$ or $2hy = \pm 2h$. When $y \neq \pm 1$, this can only happen if $h = 0$, in which case the argument has the limit $y-1/y+1$.[18] Thus the argument defines a function of h (for fixed y) which is never 0 or ∞, and is analytic in a plane cut from $h = y \pm \sqrt{y^2-1} = e^{\pm i\theta}$ (from the square root singularity in u). Since this cut plane is not simply connected, we cannot define the logarithm necessarily in the entire cut plane but must add a logarithmic cut at one of the $e^{\pm i\theta}$ points, i.e.,

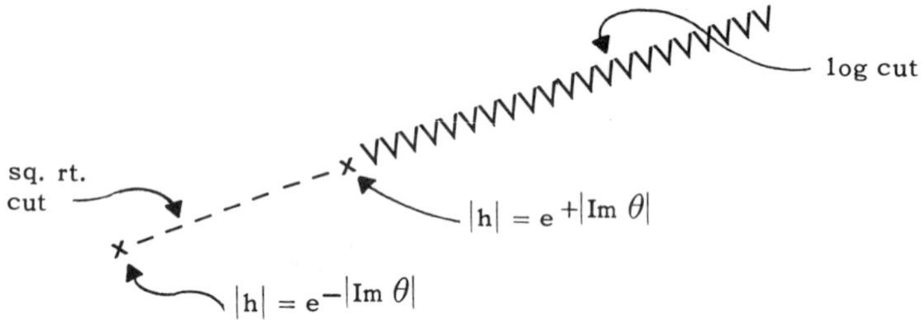

---

17 For $y = (2h)^{-1}(h^2-u^2+1)$ so that

$$\begin{aligned} G(h, y) &= \frac{1}{2u} \ln \left[\frac{(2h)^{-2}\,[h^2-u^2+1-2h^2+2hu]^2}{(2h)^{-2}\,[(h^2-u^2+1)^2-(2h)^2]} \right] \\ &= \frac{1}{2u} \ln \left[\frac{[1-(h-u)^2]^2}{[(h-1)^2-u^2][(h+1)^2-u^2]} \right] \\ &= \frac{1}{2u} \ln \left[\frac{(h-u+1)^2\,(1-h+u)^2}{(h-1+u)\,(h-1-u)\,(h+1-u)\,(h+1+u)} \right] \\ &= \frac{1}{2u} \ln \left[\frac{(h-u+1)\,(1-h+u)}{(h+u-1)\,(-1)\,(h+1+u)} \right] \end{aligned}$$

18 Either look at (VI.28) or at the definition of G where the analyticity at $h = 0$ is explicit.

Finally, we notice that the square root is not present in $G(h, y)$; for under $u \to -u$, (VI.28) is invariant (*if we are on the branch of log with log* $1 = 0$). Moreover, G doesn't have a pole at $u = 0$ on the $\log 1 = 0$ branch of log since the log vanishes there also.[19] Thus $R(y) = \max_{\pm} |e^{\pm i\theta}| = e^{|\mathrm{Im}\,\theta|}$.

The uniformity argument proceeds as in Lemma VI.20.

Remark

The singularities of $G(h, y)$ defined by $\int_{-1}^{1} dz(y-z)^{-1}(1-2hz+h^2)^{-\frac{1}{2}}$ can be found also by standard contour pinching arguments. For fixed y, the integrand has two z-plane singularities: at $z = y$ there is a fixed singularity and at $z = (h^2+1)/2h \equiv z_h$ there is a "moving singularities." There are pinches of the two singularities at $h = e^{\pm i\theta(y)}$ and possible end point singularities at $h = \pm 1$. The z-plane contour is undisturbed until

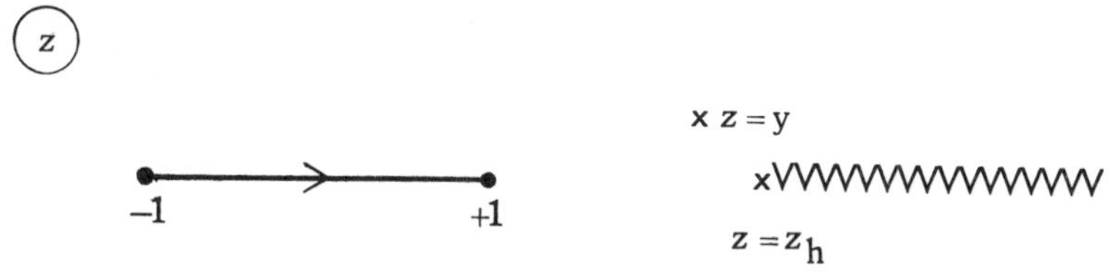

$|h| = 1$ so the pinch at $|h| = e^{-|\mathrm{Im}\,\theta|} < 1$ isn't a singularity of G since the contour isn't pinched there. However a $|h|$ goes past 1, the moving singularity "hits" the contour and starts "dragging" it along. Thus the pinch at $|h| = e^{+|\mathrm{Im}\,\theta|}$ is effective and there is a singularity there. That the end point singularities are removable is more subtle. Consider an h-plane path:

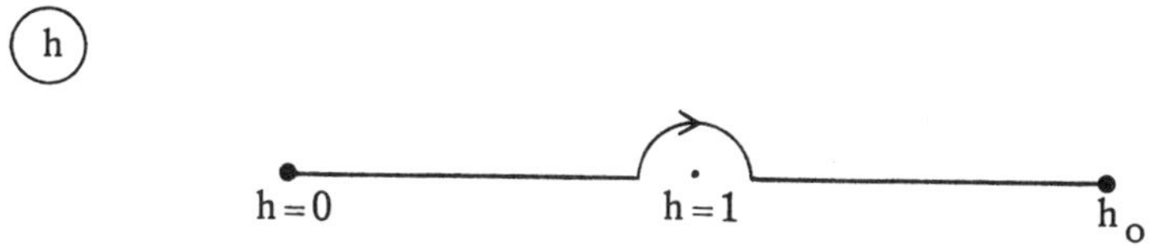

---

[19] That the singularity at $|h| = e^{-|\mathrm{Im}\,\theta|}$ isn't there is a consequence also of our definition of $G(h, y)$ since the power series, by construction, had a radius of convergence of at least 1.

At $h = 1$, $z_h = h^2 + \frac{1}{2}h$ has a vanishing derivative, so the half turn in the h-plane is a full turn in the z_h-plane; i.e., at h_o, the z-plane picture is:

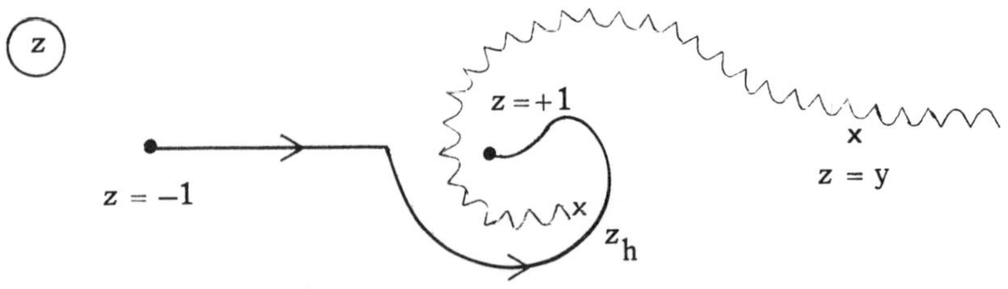

We can see the contour dragging and pinching explicitly. Now consider an h-plane path

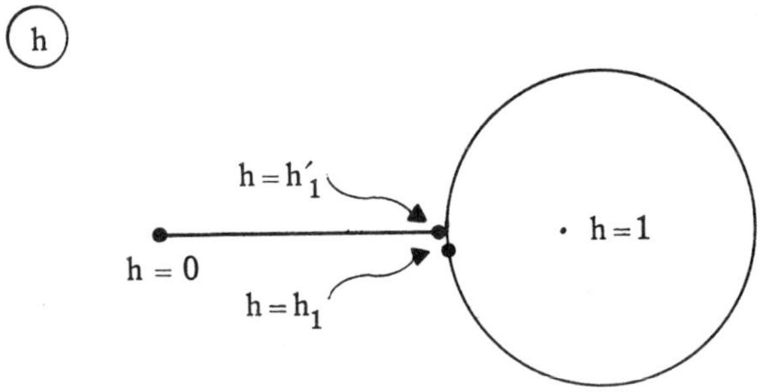

At h_1 the z-plane picture is:

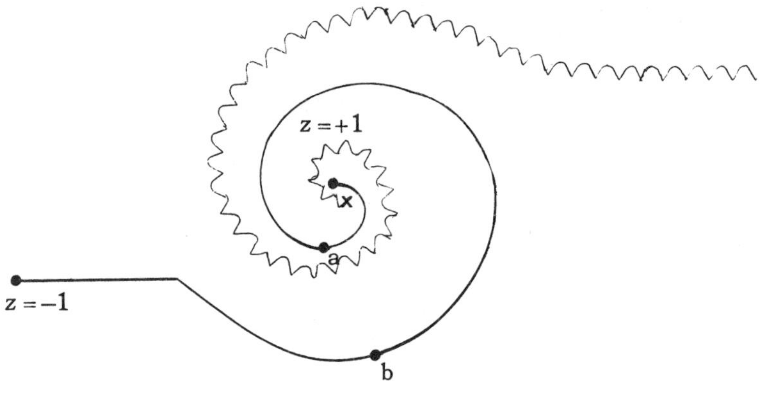

because the one turn in the h-plane makes the z_h singularity turn twice about $z = 1$. On the Rienann surface of the integrand, picture deforming the contour so that a lies over b. Because the integrand changes sign across the square root cut, the a and b curves cancel and thus $G(y, h_1) = G(y, h_1')$, i.e., the "discontinuity around the endpoint singularity" is zero so it is removable.[20]

LEMMA VI.22. *Let* z *be strictly within the ellipse drawn through* z′ *which has foci* ±1. *Then* $\sum_{\ell=0}^{\infty} (2\ell+1)\, P_\ell(z)\, Q_\ell(z')$ *converges to* $(z'-z)^{-1}$. *If* z *and* z′ *run through compacts always obeying the ellipse condition, then the convergence is uniform on the compacts.*

Proof. We have

$$\lim\, [(2\ell+1)\, P_\ell(z)\, Q_\ell(z')]^{1/\ell} \le [\lim P_\ell(z)^{1/\ell}]\, [\lim Q_\ell(z')^{1/\ell}]$$

$$= e^{|\mathrm{Im}\,\theta(z)|}\, e^{-|\mathrm{Im}\,\theta(z')|}$$

Since $|\mathrm{Im}\,\theta(z)|$ increases as the ellipse z is on increases, this is less than 1. Thus the convergence and its uniformity follow (using estimates in Lemmas VI.20 and VI.21). By Weierstrass's theorem, the limit is an analytic function f(z) for each fixed z′. Now the $P_\ell(z)$ are a complete orthonormal set on (−1, 1) with norm $\int_{-1}^{1} |P_\ell(z)|^2\, dz = \frac{2}{2\ell+1}$. Thus by definition of $Q_\ell(z')$, $\sum_{\ell=0}^{\infty} (2\ell+1)\, P_\ell(z)\, Q_\ell(z')$ converges to $(z'-z)^{-1}$ in $L^2(-1, 1)$. Therefore $f(z) = (z'-z)^{-1}$ for $z \in (-1, 1)$ and so by analytic continuation in all of the ellipse of convergence. ∎

THEOREM VI.14′. *Let* f(z) *be a function analytic in an ellipse with foci* ±1. *Then* f(z) *has a Legendre expansion:*

[20] Because the −½ power is integrable, there can't be a pole at h = 1.

$$f(z) = \sum_{\ell=0}^{\infty} (2\ell+1) a_\ell P_\ell(z) \tag{VI.21}$$

convergent uniformly on compact subsets of the open ellipse. The coefficients are given by:

$$a_\ell = \frac{1}{2} \int_{-1}^{1} dz\, f_\ell(z)\, P_\ell(z) \tag{VI.22}$$

$$= \frac{1}{2\pi i} \int_C f(z)\, Q_\ell(z)\, dz \tag{VI.23}$$

C is a contour within the cut ellipse of analyticity winding once around the cut $(-1, 1)$.

Proof. Let z be in the interior of the ellipse, E, in question. Pick a contour C within E which winds once around the ellipse through z. Then:

$$f(z) = \frac{1}{2\pi i} \int_C \frac{f(z')}{z'-z}\, dz'$$

$$= \Sigma\, (2\ell+1)\, P_\ell(z) \left(\frac{1}{2\pi i} \int_C f(z')\, Q_\ell(z')\, dz'\right)$$

by the uniformity part of Lemma VI.22. Letting $a_\ell = \frac{1}{2\pi i} \int_C f(z)\, Q_\ell(z)\, dz$, we obtain the uniform convergence of (VI.21) and formula (VI.23). The contour independence of (VI.23) is a consequence Cauchy's theorem. (VI.22) follows from the orthogonality of the P_ℓ's and the uniform convergence of (VI.21) on $(-1, 1)$. ∎

Remarks

1. The remaining parts of Theorem VI.14 (i.e., divergence outside the biggest ellipse of analyticity and $\overline{\lim}\, |a_\ell|^{1/\ell} = A + \sqrt{A^2-1}$ follow from $\lim |P_\ell(z)^{1/\ell} = e^{|\mathrm{Im}\,\theta|}$. Such an equation is a direct result of asympotic formulae for $P_\ell(z)$ as $\ell \to \infty$, formulae obtainable from suitable integral representations for $P_\ell(z)$.

2. For a detailed discussion of the analogy between Legendre and Taylor series and a method of proving results about Legendre series from theorems on Taylor series, see Section 2 of the paper of Kinoshita, Loeffel and Martin [137].

CHAPTER VII

MULTIPARTICLE SYSTEMS

VII.1. *Introduction*

We have thus far presented a more or less complete mathematical treatment of the physics of two-body quantum mechanical systems with Rollnik potentials. In this chapter, we should like to discuss some simple aspects of multiparticle systems with two body Rollnik interactions and no n-body forces ($n > 2$). We first note that:

THEOREM VII.1. *Let* $H_o = \sum_{i=1}^{N} \frac{P_i^2}{2m_i}$. *Let* $V = \sum_{i<1} V_{ij}(q_i - q_j)$ *where each* $V_{ij} \in R + L^\infty$. *Then there is a unique self-adjoint operator* H *so that* $Q(H) = Q(H_o)$ *and* $\langle\Psi, H\Psi\rangle = \langle\Psi, H_o\Psi\rangle + \langle\Psi, V\Psi\rangle$.

Proof. The uniqueness is a direct consequence of the fact that a self-adjoint operator is determined by its quadratic form. As in Chapter II, we need only show $|V_{ij}|^{1/2} \underset{T.K.}{<<} |H_o|^{1/2}$ to prove existence, or equivalently that for any a, there is a b with $\langle\Psi, V_{ij}\Psi\rangle \leq a\langle\Psi, H_o\Psi\rangle + b\langle\Psi, \Psi\rangle$ all $\Psi \in Q(H$ Suppose, $(2m_i)^{-1} \leq (2m_j)^{-1}$ and write $H_o = (2m_i)^{-1}(p_i^2 + p_j^2) + H_R$ (as an operator sum). The remainder H_R is positive, so we need only show for any a, there is a b with $\langle\Psi, V_{ij}\Psi\rangle \leq a\langle\Psi, (p_i^2 + p_j^2)\Psi\rangle + b\langle\Psi, \Psi\rangle$. But $p_i^2 + p_j^2 \geq \frac{1}{2}(p_i - p_j)^2$, so it is sufficient to prove

$$\langle\Psi, V_{ij}\Psi\rangle \leq a\langle\Psi, (p_i - p_j)^2\Psi\rangle + b\langle\Psi, \Psi\rangle,$$

for functions Ψ of n-variables. If we fix $r_1, \ldots, \hat{r}_i, \ldots, \hat{r}_j, \ldots, r_n;\ r_i+r_j$

this is just the one variable inequality used and proven in Chapter II. We can then integrate over the remaining variables. ∎

Remarks

1. Without any complication we could have allowed $H_0 = \sum_{i,j=1}^{n} a_{ij}P_iP_j$ with any positive definite matrix a_{ij},[1] since $\Sigma\, a_{ij}\, P_i\, P_j \geq C\Sigma\, P_i^2$ for some C.

2. We can of course include pure $V(\vec{r}_i)$ terms in V. One would like to study the n-body Hamiltonian defined in Theorem VII.1 in a manner analogous to our treatment of the two body Hamiltonian in Chapters III-VI. Since this has not been accomplished even in the L^2 case,[2] we do not propose to provide such a discussion here. It is our hope that the techniques we have used on the two-body case might shed some light on the n-body problem and our plan to examine this question further. We do, however, demonstrate that one of this author's favorite theorems, that of Hunziker [54] carries over to the $R + (L^\infty)_\varepsilon$ case. In order to prove Hunziker's theorem we first introduce a "factorized" Weinberg-Van Winter expansion.

VII.2. *A Factorized Weinberg-Van Winter Expansion*

In Section II.3, we replaced the "usual" equation

$$(E+H)^{-1} = [1+(E+H_0)^{-1}V]^{-1}\,(E+H_0)^{-1}$$

with a "factorized" Tiktopoulos formula

$$(E+H)^{-1} = (E+H_0)^{-1/2}[1+(E+H_0)^{1/2}\,V(E+H_0)^{-1/2}]^{-1}\,(E+H_0)^{-1/2}.$$

---

1 This is not quite pure pedantry. For when the center of mass is removed, a_{ij} is not diagonal; the so called Hughes-Eckart [138] or specific mass corrections appear as $P_i \cdot P_j$ terms (see Appendix 1 to VII.3).

2 For a discussion of what is known in this case, see Hepp [49], Hunziker [55] and Kato [71]. Added in proof: there has been an important breakthrough in the study of n-body systems while this monograph was in press. It is due to E. Balslev and J. M. Combes and is described in a SUNY at Buffalo preprint.

This was done to obtain the Hilbert-Schmidt operator $(E+H_o)^{-1/2}\, V(E+H_o)^{-1/2}$ in place of the not-well-defined-Hilbert-space-operator $(E+H_o)^{-1}V$ in case $V \in R + L^{\infty}$. In this section, it is our goal to factorize the multiparticle resolvent equation of Weinberg [117] and Van Winter [124] in the same way.

For simplicity of presentation, we consider first the equal-mass three body case. We recall Weinberg's motivation for the introduction of his expansion. If one considers the Green's function equation:

$$G(E) = G_o(E) + \tilde{I}(E)G(E) \tag{VII.1a}$$

where

$$G(E) = (E-H)^{-1} \tag{VII.1b}$$

$$G_o(E) = (E-H_o)^{-1} \tag{VII.1c}$$

$$\tilde{I}(E) = (E-H_o)^{-1}\ (V_{12} + V_{23} + V_{13}) \tag{VII.1d}$$

one sees that any attempt to treat (VII.1a) directly using the theory of Fredholm integral equations is doomed to failure. This is basically because $(E-H_o)^{-1}\, V_{12}$ when written as a kernel[3] $k(\vec{r}_{12}, \vec{r}_{23}; \vec{r}\,'_{12}, \vec{r}\,'_{23})$, depends on $2r_{23} + r_{12}$, $2r'_{23} + r'_{12}$ only through their difference.[4] Put differently, the Fourier transform of K has a $\delta(\vec{P}_3 - \vec{P}'_3)$ since 3 undergoes no interactions. The rub then is that to get a nice kernel we must have interactions between all the particles in any term of the kernel so $G_o(E)$ is not the proper inhomogeneous term. We must include terms like $(E-H_o)^{-1}\, V_{12}\, (E-H_o)^{-1}$ in the inhomogeneous part of (VII.2a).

---

3 We have "removed" the center of mass $\vec{R}$ from K; if we hadn't, K would contain a $\delta(\vec{R}-\vec{R}'))$. Alternatively we have $[(E-H_o)^{-1}\, V_{12}\Psi]\, (\vec{r}_{12}, \vec{r}_{23}, \vec{R}) = \int d^3r'_{12}\, d^3r'_{23}\, K(\vec{r}_{12}, \vec{r}_{23}; \vec{r}\,'_{12}\, \vec{r}\,'_{23})\, \Psi(\vec{r}\,'_{12}, \vec{r}\,'_{23}, \vec{R})$. Since G(E) also has the cente of mass δ-function, its presence in $\tilde{I}$ causes no problem, since it factors out of the equation.

4 This combination comes from $\vec{r}_3 = \vec{R} - \frac{1}{3}\,(2\vec{r}_{23} + \vec{r}_{12})$ where $\vec{R} = \frac{1}{3}\, \vec{r}_1 + \vec{r}_2 + \vec{r}_3$ and $\vec{r}_{ij} = \vec{r}_i - \vec{r}_j$.

One thus iterates (VII.1a):[5]

$$G(E) = G_0(E) + G_0(E) \sum_{i<j} V_{ij} G_0(E) + G_0(E) \sum_{i<j} V_{ij} G_0(E) \sum_{i<j} V_{ij} G_0(E) + \dots \tag{VII.2}$$

and introduces (following Weinberg) the graphical symbolism:

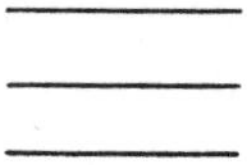
$\equiv G_o$;

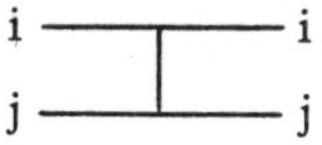

$\equiv V_{ij}$

so that, for example:

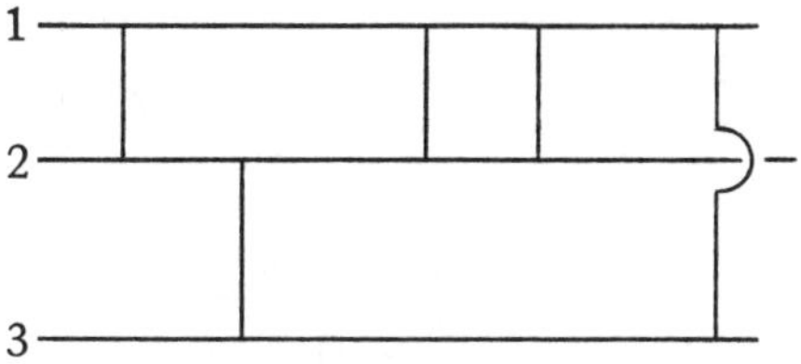

$= G_0V_{12}G_0V_{23}G_0V_{12}G_0V_{12}G_0V_{13}G_0$

Thus letting

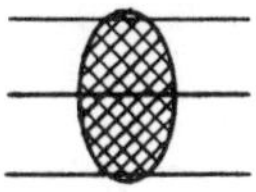
$\equiv G$

(VII.2) becomes:

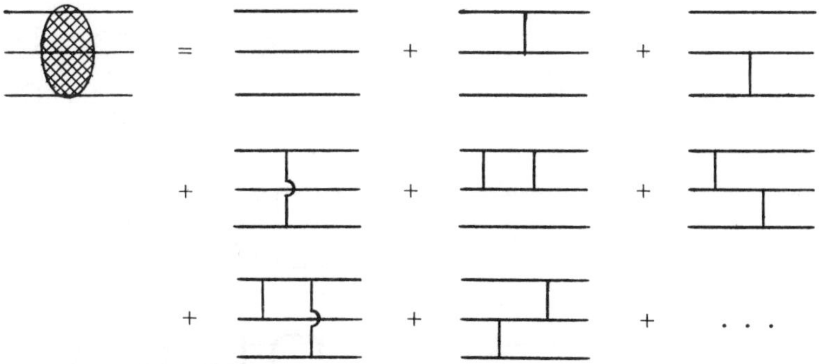

5 We consider the convergence of this iteration below.

The terms which would present momentum space delta functions if we put them into I are precisely the *disconnected* diagrams. Weinberg thus writes

$$G(E) = D(E) + I(E)\,G(E) \tag{VII.3a}$$

with D ≡ sum of disconnected diagrams. By resumming diagrams, we can be more explicit. Let

$$G^{(ij)} \equiv (E - H_o - V_{ij})^{-1} \tag{VII.4}$$

be a two particle Green's function and write:

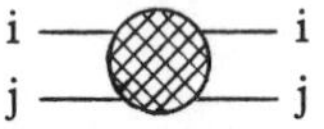

$\equiv G^{(ij)}$

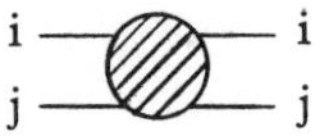

$= G_o V_{ij} G^{(ij)} = G^{(ij)} - G_o$

i.e., 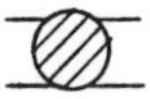 is the connected part of 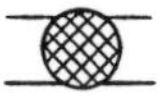. Then, one has (VII.3a) in resummed form:

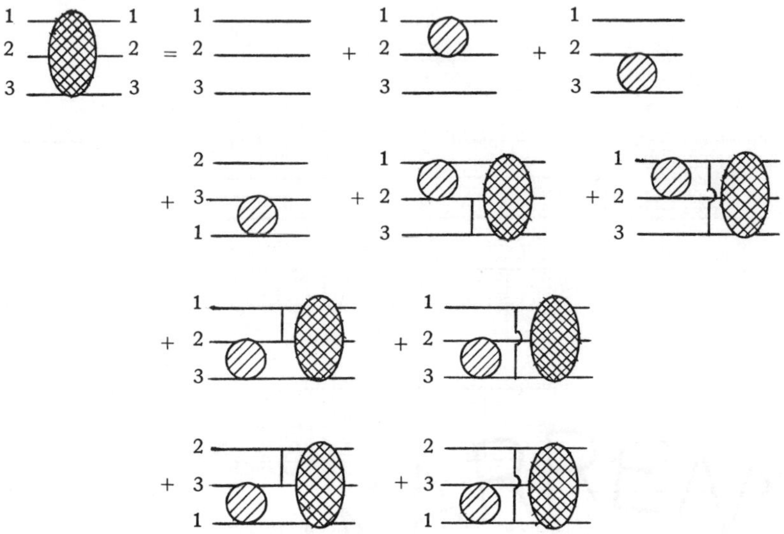

The first terms are D, i.e.,

$$D = G_o + \sum_{i<j} G_o V_{ij} G^{(ij)} \tag{VII.3b}$$

$$I = \sum_{i<j} G_o V_{ij} G^{(ij)} (V_{ik} + V_{jk}) \quad (k \neq i, j). \tag{VII.3c}$$

That (VIII.3) holds is a purely "diagramatic" argument. This completes the review of the formal equations in the L^2-case.

In order to obtain a formula analogous to the Tiktopoulos formula, we symmetrize all things or else "reduce" out a $G_o^{1/2}$ from both ends.[6] Explicitly, define the reduced Green's functions:

$$R^{(ij)} = (1 - G_o^{1/2} V_{ij} G_o^{1/2})^{-1} \tag{VII.5a}$$

$$R = (1 - G_o^{1/2} \sum_{i<j} V_{ij} G_o^{1/2})^{-1} \tag{VII.5b}$$

the reduced disconnected part:

$$D_R = 1 + \sum_{i<j} G_o^{1/2} V_{ij} G_o^{1/2} R^{(ij)} \tag{VII.5c}$$

and the symmetrized connected interaction:

$$I_S = \sum_{i<j} G_o^{1/2} V_{ij} G_o^{1/2} R^{(ij)} G_o^{1/2} (V_{ik} + V_{jk}) G_o^{1/2} \quad (k \neq i, j) \tag{VII.5d}$$

Formally:

$$R = D_R + I_S R \tag{VII.6}$$

so that we are led to a symmetrized Weinberg-Van Winter formula:

---

6 By $G_o^{1/2}$ we mean multiplication by $(E - k^2)^{-1/2}$ in momentum space chosen with some particular branch of square root, say $\mathrm{Im}(-|E|-k^2)^{-1/2} > 0$.

THEOREM VII.2 (Three particle symmetrized Weinberg-Van Winter formula). *Let* $H = H_o + \sum_{i<j} V_{ij}$ *be defined by a forms method where each* $V_{ij} \in R + L^\infty$. *Then for* E *sufficiently negative, or for* E *in the cut plane with sufficiently negative real part:*

$$(E-H)^{-1} = (E-H_o)^{-\frac{1}{2}} (1-I_S)^{-1} D_R (E-H_o)^{-\frac{1}{2}}. \qquad \text{(VII.7)}$$

Proof. If we regard G_o. G^{ij} as maps of $\mathcal{H}_{-1}$ to $\mathcal{H}_{+1}$ and the V_{ij} as maps of $\mathcal{H}_{+1}$ to $\mathcal{H}_{-1}$, the infinite series (VII.2) converges for E sufficiently negative (all one needs is $\|V(E-H_o)^{-1}\| \leq 1$). (VII.6) then follows by resummation and the "peeling" of $(E-H_o)^{-\frac{1}{2}}$ off both ends. We thus obtain $R = (1-I_S)^{-1} D_R$ when E is so negative that we know $\|I_S\| < 1$. Thus using $(E-H)^{-1} = (E-H_o)^{-\frac{1}{2}} R(E-H_o)^{-\frac{1}{2}}$ our proof is complete. ∎

Remark

We will see that $I_S(E)$ is analytic and compact if $E \notin \text{spec}(H_{ij})$ for any ij ($H_{ij} = H_o + V_{ij}$) if $V_{ij} \in R + (L^\infty)_\varepsilon$. From this (VII.7) will extend to all $E \notin \text{spec}(H_{ij})$ except for a discrete set at worst. This discrete set will contain any eigenvalues of H.

To discuss the general n-particle Weinberg-Van Winter equations, we adopt the (notational) formalism of Hunziker [53, 54, 55]. We first make several definitions:

DEFINITION. A *cluster decomposition* is a decomposition, D, of $\{1,\ldots,n\}$, i.e., a family of disjoint subsets of $\{1,\ldots, n\}$ whose union is $\{1,\ldots, n\}$. Usually this is called a partition of $\{1,\ldots, n\}$).

We think of D as a partition of the particles $1,\ldots, n$ into clusters. We automatically associate with D, several operators related to the Hamiltonian with the interaction between clusters turned off.

DEFINITION. Given a cluster decomposition D we write iDj if i and j lie in the same element of D[7] and ~iDj if i and j lie in different elements of D.

DEFINITION. Given a cluster decomposition, D, and an n particle Hamiltonian $H = H_o + \sum_{i<j} V_{ij}$ we define:

$$H_D = H_o + \sum_{i<j;\ iDj} V_{ij} \tag{VII.7}$$

$$V_D = \sum_{i<j;\ iDj} V_{ij} \tag{VII.8}$$

$$I_D = \sum_{i<j;\ \sim iDj} V_{ij} \tag{VII.9}$$

We call these objects the cluster Hamiltonian, the cluster potential and the (inter-) cluster interaction. Obviously, $H_D = H_o + V_D$ and $H_D = H - I_D$.

DEFINITION. Given two cluster decompositions D_1 and D_2, we say D_2 is a *refinement of* D_1 and write $D_1 \lhd D_2$ if D_2 is obtained by further partitioning, i.e., if for every $A \in D_2$, there is a $B \in D_1$ with $A \subset B$.

DEFINITION. A *string*, S, is an ordered family of cluster decompositions $S = (D_n, D_{n-1}, \ldots, D_k)$ with $D_n \rhd D_{n-1} \rhd \ldots \rhd D_k$ so that:

(a) $D_n = \{\{1\}, \{2\}, \ldots, \{n\}\}$

(b) D_ℓ has ℓ subsets.

Thus $D_{\ell-1}$ is obtained from D_ℓ by joining two elements of D_ℓ together.

Strings enter because Weinberg graphs have associated strings, as follows: Consider a graph and cut it with a vertical line, L, at the ex-

---

7 Recall D is a family of subsets.

treme left. To the left of L, the graph breaks up into pieces $D_n = \{\{1\},\ldots, \{n$ Move L to the right. As we cross each interaction, the set of pieces to the left of L either changes by joining two components together or it doesn't change. A string thereby results. For example consider:

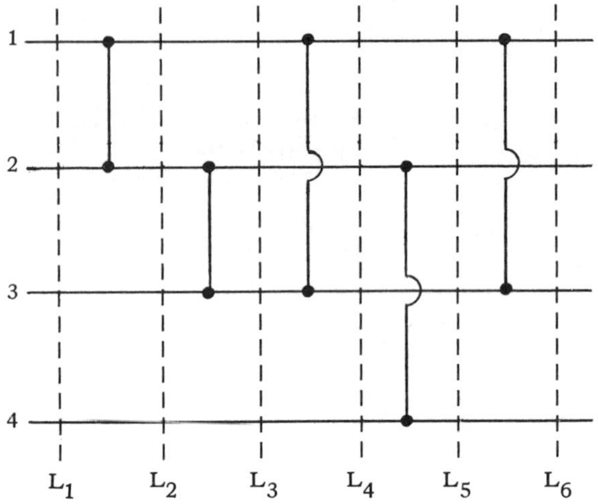

The associated string is $\{D_4, D_3, D_2, D_1\}$ where $D_4 = \{\{1\}, \{2\}, \{3\}, \{4\}\}$; $D_3 = \{\{1,2\}, \{3\}, \{4\}\}$, $D_2 = \{\{1, 2, 3\}, \{4\}\}$; $D_1 = \{\{1, 2, 3, 4\}\}$. Notice in shifting from L_3 to L_4 or from L_5 to L_6, D doesn't change.

DEFINITION. We say a string is connected if the right most element of S is $\{\{1,\ldots, n\}\} = D_1$.

Thus a graph is connected if and only if its associated string is connected. One final definition is needed before we return to the Weinberg expansion:

DEFINITION. We say a graph has *associated cluster decomposition* D(G), if D(G) is the right most member of S(G). We say G is D_k-*disconnected* if $D_k \lhd D(G)$.

Now, analogously to (VII.1) and (VII.4), we define

$$G(E) = (E - H)^{-1} \qquad \text{(VII.10a)}$$

$$G_0(E) = (E - H_0)^{-1} \qquad \text{(VII.10b)}$$

$$G_D(E) = (E - H_D)^{-1} \qquad \text{(VII.10c)}$$

Then a simple analysis shows that

$$G_D(E) = \Sigma \text{ all D-disconnected diagrams.} \qquad \text{(VII.11)}$$

We can write G(E) as a sum over all diagrams, but all disconnected diagrams are not $\sum_{D \text{ disconnected}} G_D$ since this multiple-counts some diagrams. Even in the three particle case, we saw that to avoid multiple counting of , we set $D = G_0 + \Sigma[G^{(ij)} - G_0]$. While every diagram is not D-disconnected for a unique D, every diagram has a unique associated string S. We thus define:

$$G_S(E) = \Sigma \text{ all diagrams with } S(G) = S. \qquad \text{(VII.12)}$$

If $S = (D_N, D_{N-1}, \ldots D_k)$, a simple diagramatic analysis shows:

$$G_S(E) = G_{D_N}(E) V_{D_N D_{N-1}} G_{D_{N-1}}(E) \ldots V_{D_{k+1} D_k} G_{D_k}(E) \qquad \text{(VII.13a)}$$

where

$$V_{D_\ell D_{\ell-1}} = I_{D_\ell} - I_{D_{\ell-1}} = V_{D_{\ell-1}} - V_{D_\ell}. \qquad \text{(VII.13b)}$$

Thus in the example above:

$$V_{D_3 D_2} = V_{13} + V_{23}.$$

Now analogously to (VII.3), we have (in the L^2 case)

$$G(E) = D(E) + I(E)G(E) \qquad \text{(VII.14a)}$$

where

$$D = \sum_{S \text{ disconnected}} G_S(E) \tag{VII.14b}$$

and

$$I = \sum_{S=\{D_N,\ldots,D_1\}} G_{D_N} V_{D_N D_{N-1}} G_{D_{N-1}} \cdots G_{D_2} V_{D_2 D_1} \;. \tag{VII.14c}$$

Analogously to (VIII.5) we define the reduced Green's functions:

$$R_D = (1 - G_o^{1/2} V_D G_o^{1/2})^{-1} \tag{VII.15a}$$

$$R = (1 - G_o^{1/2} V G_o^{1/2})^{-1} \tag{VII.15b}$$

and the reduced disconnected part

$$D_R = \sum_{S \text{ disconnected}} R_{D_N} (G_o^{1/2} V_{D_N D_{N-1}} G_o^{1/2}) R_{D_{N-1}} \cdots$$

$$(G_o^{1/2} V_{D_{k+1} D_k} G_o^{1/2}) R_{D_k} \;. \tag{VII.15c}$$

(Note $R_{D_N} = 1$)

and the "symmetrized" connected interaction:

$$I_S = \sum_{S \text{ connected}} (G_o^{1/2} V_{D_N D_{N-1}} G_o^{1/2}) R_{D_{N-1}} \cdots$$

$$R_{D_2} (G_o^{1/2} V_{D_2 D_0} G_o^{1/2}) \;. \tag{VII.15d}$$

Then, proceeding in the same manner as in the proof of Theorem VII.2, we obtain:

THEOREM VII.3 (N-particle Weinberg Van-Winter formula). *Let* $H = H_o + \sum_{i<j} V_{ij}$ *be defined by a forms method where each* $V_{ij} \in R + L^\infty$. *Then for* E *sufficiently negative, or for* E *in the cut plane with sufficient negative real part:*

$$(E-H)^{-1} = (E-H_o)^{-1/2} (1-I_S)^{-1} D_R (E-H_o)^{-1/2} . \qquad \text{(VII.16)} \blacksquare$$

Finally, we extend the result of this section to a slightly more general Hamiltonian. First we can allow $H_o = \Sigma a_{ij} p_i p_j$ where a_{ij} is a positive definite form. Secondly, we can include potentials $V_i(\vec{r}_i)$. In this case we add a fictitious line "0" and write $\vec{r}_i = \vec{r}_{io}$, $V_i = V_{io}$ and define connected, etc., in terms of this line 0.

The usual Hamiltonian, $\Sigma(2m_i)^{-1} P_i^2 + \sum_{i<j} V_{ij}(r_{ij})$, is translation invariant and this is reflected in two places: (a) The Green's functions have $\delta(R-R')$ terms as we have noted; (b) Bound states aren't eigenfunctions of H but rather of $H - H_{C.M.}$. It is convenient thus to change coordinates and eliminate the center of mass. One way of explicitly accomplishing this is to single out $\vec{r}_n$ and introduce new coordinates:

$$\vec{r}_i{}' = \vec{r}_i - \vec{r}_n \quad (i = 1, \ldots, n-1) \qquad R = (\sum_{i=1}^{n} m_i)^{-1} (\sum_{i=1}^{n} m_i \vec{r}_i)$$

Then, $H = \Sigma(2m_i)^{-1} P_i^2 + \sum_{i<j} V_{ij}(r_{ij})$ becomes $H = (2\Sigma m_i)^{-1} P^2 + \sum_{i,j<n} a_{ij} \vec{p}_i{}' \vec{p}_j{}' + \sum_{i<j<n} V_{ij}(r_{ij}{}') + \sum_{i<n} V_{in}(\vec{r}_i{}')$. If the center of mass coordinate is "removed," we have precisely a "generalized problem" of the type considered above.

VII.3. *Hunziker's Theorem*

In [54], Hunziker proved the following:[8]

[8] This generalized a result proven earlier by Van Winter [124].

THEOREM VII.4. *Let* $H = H_o + \sum_{i<j} V_{ij}(r_{ij})$ *with* $H_o = \Sigma(2m_i)^{-1} P_i^2$ *and each* $V_{ij} \in L^2 + (L^\infty)_\varepsilon$. *Let* H_D *be the cluster Hamiltonian given by* (VII.7). *Let* $C = \min_{D \neq \{1,\dots n\}} \inf \sigma(H_D)$. *Then* $\tilde{H}$, *i.e.,* H *with the center of mass removed, has only discrete spectrum below* C.

Remarks

1. Hunziker also showed that $\sigma(\tilde{H})$ is essential in $[C, \infty)$ so C is the bottom of the "continuum."

2. From a physical point of view, this theorem is entirely reasonable, in fact, it is down-right pleasing! It says that continuum states come from non-interacting clusters arbitrarily far from one another.

3. Hunziker is not very explicit in his discussion of removing the center of mass. We return to this question in Appendix 1 to this section.

4. In Appendix 2 we show how this theorem directly implies that the Helium atom has an infinite number of bound states.

Hunziker's proof, which we intend to mimic (more or less) is a three step affair:

I. When $V \in L^2$, one proves $I(E)$ is Hilbert-Schmidt if E is in the plane cut from C to ∞ [53, 124].

II. By a limiting argument one shows $I(E)$ is compact in general.

III. By using the Analytic Fredholm Theorem (Theorem A.27), the proof is completed.

We generalize Hunziker's theorem by proving:

THEOREM VII.5. *Let* H, H_o, H_D, C, $\tilde{H}$ *be as in Theorem* VII.4 *except suppose only that* $V_{ij} \in R + (L^\infty)_\varepsilon$ *and let* H, H_D *be defined by the forms method. Then* $\tilde{H}$ *has only discrete spectrum in* $(-\infty, C)$.

Remarks

1. This is a strict generalization of Hunziker's theorem since

$$L^2 + (L^\infty)_\varepsilon \subset R + (L^\infty)_\varepsilon .$$

2. We suppose we have removed the center of mass and have

$$H = \sum_{i,j=1,\ldots,N-1} a_{ij}P_iP_j + \sum_{i<j<N} V_{ij}(r_{ij}) + \sum_{i<N} V_{io}(r_i)$$

Proof. I. $I_S(E)$, *given by* (VII.15d), *is Hilbert-Schmidt if the* $V_{ij} \in R$ *and* E *is in the plane cut from* C *to* ∞. We consider each individual term of I_S separately, i.e., a general term

$$T = (G_o^{1/2}V_{i_Nj_N}G_o^{1/2})R_{D_{N-1}}(G_o^{1/2}V_{i_{N-1}j_{N-1}}G_o^{1/2}) \cdots (G_o^{1/2}V_{i_2j_2}G_o^{1/2})$$

where i_r, j_r are indices in different clusters of D_r but the same cluster of D_{r-1}.

Consider the coordinates $\vec{r}_2{}', \ldots, \vec{r}_N{}'$

$$\vec{r}_k{}' = \vec{r}_{i_kj_k} \tag{VII.17}$$

where $\vec{r}_{io} = \vec{r}_i$. We first claim that the $\vec{r}_k{}'(k = 2,\ldots, N)$ are a complete set of coordinates if the term under consideration comes from a connected graph (as we are assuming). For setting $\vec{r}_{ij} = 0$ ties together $\vec{r}_i$ and $\vec{r}_j$ unless $j = 0$ in which case $r_i = 0$. To say the graph is connected is to say if we tie together all pairs connected by vertical lines all the lines 0, 1,..., N–1 are in one bundle. Thus $\vec{r}_i{}' = 0$, $i = 2,\ldots, N$ implies $\vec{r}_1 = \vec{r}_2 = \ldots = \vec{r}_{N-1} = 0$. This says the coordinates $\vec{r}_i{}'$ are a complete (and by their number) independent set.

We deal with the kernel, $\mathcal{T}$, of T in momentum space and prove $\mathcal{T}(P_2{}',\ldots, P_N{}'; \tilde{P}_2{}',\ldots, \tilde{P}_N{}')$ is in $L^2(P', \tilde{P}')$. Following Hunziker [53], we note that $T_r = (G_o^{1/2}V_{i_Nj_N}G_o^{1/2}) R_{D_{N-1}} \ldots R_{D_r}$ has no forces between clusters of D_r, so it has a kernel with $\delta(P_2{}' - \tilde{P}_2{}') \ldots \delta(P_r{}' - \tilde{P}_r{}')$, that is

$$(T_r f)(P_2',\ldots,P_N') = \int d\tilde{P}'_{r+1} \cdots d\tilde{P}_N' \, \mathcal{T}^{(P_2',\ldots,P_r')}(P_{r+1}',\ldots,P_N'; \tilde{P}_{r+1}',\ldots,\tilde{P}_N') \, f(P_2',\ldots,P_r', \tilde{P}_{r+1}',\ldots,\tilde{P}_N') \quad \text{(VII.18)}$$

We will show inductively that $\mathcal{T}^{(P_2',\ldots,P_r')} \in L^2(P_{r+1}',\ldots,P_N'; \tilde{P}_{r+1}',\ldots,\tilde{P}_N')$ for all $P_2',\ldots,P_r'$ and has L^2 norm bounded by a constant independent of $(P_2',\ldots,P_r')$. These statements all have a mathematically natural interpretation in the realization

$$L^2(P_1',\ldots,P_N') = L^2(P_1',\ldots,P_r') \otimes L^2(P_{r+1}',\ldots,P_N')$$

and $L^2(\Omega) \otimes M = L^2(\Omega; M)$ the functions from Ω to M (i.e., $\int_\Omega^\oplus M d\mu$). We describe this in detail in Appendix 3 to this section. Here we will use the short hand notation of δ-function kernels more familiar to physicists.

Start the induction with $r = N$ so $T_r = 1 = \mathcal{T}^{(P_2',\ldots,P_N')}$. Clearly $\mathcal{T}^{(P_2',\ldots,P_N')} \in L^2$ (point) with L^2 norm $(=|\mathcal{T}|^2 = 1)$ independent of $\vec{P}'$. Suppose we know T_{r+1} obeys our claim. Let $S_r = T_{r+1}(G_0^{1/2} V_{r+1} G_0^{1/2})$. Then S_r has "kernel":

$$\mathcal{S}_r^{(P_1',\ldots,P_{r-1}')}(P_r',\ldots,P_N'; \tilde{P}_r',\ldots,\tilde{P}_N')$$

$$= T_{r+1}^{(P_1',\ldots,P_r')}(P_{r+1}',\ldots,P_N',\ldots,P_N'; \tilde{P}_{r+1},\ldots,\tilde{P}_N')$$

$$G_0^{1/2}(P_1',\ldots,P_r', \tilde{P}_{r+1}',\ldots,\tilde{P}_N')$$

$$\hat{V}_r(P_r' - \tilde{P}_r') G_0^{1/2}(P_1',\ldots,P_{r-1}', \tilde{P}_r',\ldots,\tilde{P}_N') \quad \text{(VII.19)}$$

where V_r is the Fourier transform of V_r (suitably normalized!) and

$G_0^{1/2}(P_1',\ldots,P_N') = i(\Sigma a_{ij}'P_i'P_j' + E)^{-1/2}$ where a_{ij}' is a suitable positive definite matrix. Since a_{ij}' is positive definite we can find C with $\Sigma a_{ij}'P_i'P_j' \geq C\Sigma P_i'^2$. Thus $|G_0^{1/2}(P_1',\ldots,P_N')| \leq |CP_r'^2 + E|^{-1/2}$. As a result

$$|\mathcal{S}_r^{(P_1',\ldots,P_{r-1}')}(P_r',\ldots,P_N';\tilde{P}_r',\ldots,\tilde{P}_N')|^2$$

$$\leq |T_{r+1}^{(P_1',\ldots,P_r')}(P_{r+1}',\ldots,P_N';\tilde{P}_{r+1}',\ldots,\tilde{P}_N')|^2$$

$$|(CP_r'^2 + E)^{-1/2}\,\hat{V}(P_r' - P_r')(CP_r'^2 + E)^{-1/2}| \qquad \text{(VII.20)}$$

The $|T_{r+1}|^2$ is integrable over $(P_{r+1}',\ldots,P_N',\tilde{P}_{r-1}',\ldots,\tilde{P}_N')$, independent of $(P_1',\ldots,P_r')$. By (I.15), the $|C\ldots\hat{V}\ldots C\ldots|$ term is in $L^2(P_r',\tilde{P}_r')$ so $\mathcal{S}_r \in L^2(P_r',\ldots,P_N';\tilde{P}_r',\ldots,\tilde{P}_N')$ with L^2 norm bounded independently of $(P_1',\ldots,P_{r-1}')$. Now $T_r = S_r D_r$. D_r has $\delta(P_1' - \tilde{P}_1')\ldots\delta(P_{r-1}' - \tilde{P}_{r-1}')$ in its kernel (since $r_1',\ldots,r_{r-1}'$ are intercluster coordinates) and thus $D_r = \mathcal{D}_R^{(P_1',\ldots,P_{r-1}')}$ a family of uniformly bounded operators on $L^2(P_r',\ldots,P_N')$[9] (uniform in $(P_1',\ldots,P_{r-1}')$. Thus

$$\mathcal{T}_r^{(P_1',\ldots,P_{r-1}')} = \mathcal{S}_r^{(P_1',\ldots,P_{r-1}')}\mathcal{D}_r^{(P_1',\ldots,P_{r-1}')}$$

is Hilbert-Schmidt as the product of a bounded and a Hilbert-Schmidt operator (on $L^2(P_r',\ldots,P_N')$) with uniform H–S norm. This completes the induction proof.

---

9 This is particularly clear in the $L^2(\Omega, M)$ formulation; see Appendix 3 to this section.

II. $I_S(E)$ *given by* (VII.15d), *is compact if the* $V_{ij} \in R + L^\infty)_\varepsilon$ *and* E *is the plane cut from* C *to* ∞. For $V_{ij} \in R + (L^\infty)_\varepsilon$ means one can find $V_{ij}^{(n)} \in R$ with $\|V_{ij} - V_{ij}^{(n)}\|_\infty < \frac{1}{n}$. It is easy to see that in this case the $I_S^{(n)}(E)$ associated with the $V_{ij}^{(n)}$ converge to $I_S(E)$ in norm for any fixed E in the cut plane.[10] As a limit of Hilbert-Schmidt operators, $I_S(E)$ is compact.

III. *Completion of the Proof.* It is not hard to see $I_S(E)$ is analytic for E in the cut plane and $\|I_S(E)\| \to 0$ as $E \to -\infty$. Thus by the analytic Fredholm theorem, $(1-I_S(E))^{-1}$ exists in the cut plane, except on a discrete set. By Theorem V88.3, $(E - H)^{-1}$ exists in the cut plane, except on a discrete set which completes the proof.

Appendix 1 to VII.3. Removing the Center of Mass

For the reader's convenience we summarize how Hughes-Eckart terms arise and we explain why they don't affect Hunziker's theorem. For simplicity, consider first, the Helium atom Hamiltonian.

$$\frac{1}{2m}\vec{P}_1^{\,2} + \frac{1}{2m}\vec{P}_2^{\,2} + \frac{1}{2m}\vec{P}_3^{\,2} - \frac{2e^2}{|\vec{r}_2 - \vec{r}_1|} - \frac{2e^2}{|\vec{r}_3 - \vec{r}_1|} + \frac{e^2}{|\vec{r}_3 - \vec{r}_2|} = H_{He} \qquad \text{(VII.21)}$$

It is more or less natural to introduce the coordinates:

$$\vec{R} = (M\vec{r}_1 + m\vec{r}_2 + m\vec{r}_3)(M + 2m)^{-1} \qquad \text{(VII.22a)}$$

$$\vec{r}_{12} = \vec{r}_2 - \vec{r}_1 \qquad \text{(VII.22b)}$$

$$\vec{r}_{13} = \vec{r}_3 - \vec{r}_1 \qquad \text{(VII.22c)}$$

---

10 One must first show $(H_D^{(n)} + E)^{-1} \to (H_D + E)^{-1}$ for E large to know any E in the cut plane is eventually outside spec $(H_D^{(n)})$ for all D.

In general, we consider $H_o = \sum_{i=1}^{n} (2m_i)^{-1} P_i^2$ and set

$$M = \sum_{i=1}^{n} m_i \tag{VII.23a}$$

$$R = M^{-1} \sum_{i=1}^{n} m_i\vec{r}_i \tag{VII.23b}$$

$$\vec{\eta}_i = \vec{r}_i - \vec{r}_n \qquad i = 1,\dots, n-1. \tag{VII.23c}$$

From $\sum_{i=1}^{n-1} m_i\eta_i = \sum_{i=1}^{n} m_i\vec{r}_i - \sum_{i=1}^{n} m_i\vec{r}_n = M\vec{R} - M\vec{r}_n$, we conclude

$$\vec{r}_n = \vec{R} - \sum_{i=1}^{n-1} M^{-1} m_i\vec{\eta}_i \tag{VII.23d}$$

$$\vec{r}_i = \vec{R} + \vec{\eta}_i - \sum_{i=1}^{n-1} M^{-1} m_i\vec{\eta}_i \qquad i = 1,\dots, n-1. \tag{VII.23e}$$

Thus, letting:

$$\zeta = \sum_{i=1}^{n-1} M^{-1} m_i\eta_i : \tag{VII.24}$$

$$2H_o = \sum_{i=1}^{n} m_i\dot{r}_i^2$$

$$= \sum_{i=1}^{n-1} m_i(\dot{R} + \dot{\eta}_i - \dot{\zeta})^2 + m_n(\dot{R} - \dot{\zeta})^2$$

so:

$$2H_o = M\dot{R}^2 + \sum_{i=1}^{n-1} m_i\dot{\eta}_i^2 - M\dot{\zeta}^2 \tag{VII.25a}$$

$$[= M\dot{R}^2 + \sum_{i=1}^{n-1} m_i(1-M^{-1}m_i)\dot{\eta}_i^2 - \sum_{i\neq j} M^{-1}m_im_j\, \dot{\eta}_i\cdot\dot{\eta}_j] \tag{VII.25b}$$

Let P_i be conjugate to η_i, i.e., $P_i = \partial T/\partial \dot{\eta}_i$ and P conjugate to R. Since $\partial\dot{\zeta}/\partial\dot{\eta}_i = M^{-1}m_i$:

$$P = M\dot{R} \tag{VII.26a}$$

$$P_i = m_i\dot{\eta}_i - m_i\dot{\zeta} \tag{VII.26b}$$

Thus

$$\dot{\eta}_i = m_i^{-1} P_i + \dot{\zeta} \tag{VII.27}$$

so in particular $\dot{\zeta} = \sum_{i=1}^{n-1} M^{-1}m_i\dot{\eta}_i = \sum_{i=1}^{n-1} M^{-1} P_i + (1-Mm_n^{-1})\dot{\zeta}$ or

$\dot{\zeta} = m_n^{-1} \sum_{i=1}^{n-1} P_i$. Thus

$$2H_o = M^{-1}P^2 + \sum_{i=1}^{n-1} (m_i^{-1} P_i^2 + 2\dot{\zeta}\cdot P_i) + \dot{\zeta}^2(-M + \sum_{i=1}^{n} m_i)$$

$$= M^{-1}P^2 + \sum_{i=1}^{n-1} m_i^{-1}P_2^2 + \dot{\zeta}\cdot P_i.$$

$$H_o = \frac{P^2}{2M} + \sum_{i=1}^{n-1} \frac{\vec{P}_i^{\,2}}{2\mu_i} + \sum_{i\neq j\leq n-1} \frac{\vec{P}_i\cdot\vec{P}_j}{2m_n} \tag{VII.28}$$

where:

$$\mu_i^{-1} = m_i^{-1} + m_n^{-1}. \tag{VII.29}$$

For the Helium atom, letting $\mu = Mm/M+m$ we see

$$H_{He} - H_{C.M.} = \frac{1}{2\mu}(\vec{P}_{12}^{\,2} + \vec{P}_{13}^{\,2}) - \frac{2e^2}{|r_{12}|} - \frac{2e^2}{|r_{13}|} \tag{VII.30}$$

$$+ \frac{e^2}{|\vec{r}_{12} - \vec{r}_{13}|} + \frac{1}{M}\vec{P}_{12}\cdot\vec{P}_{13}.$$

One feels that the continuum for the Helium atom should begin at the binding energy of the He^+ ion. At first sight this does not *appear* to be the case, for Hunziker's theorem tells us that the continuum begins at $C = \min(C_{12}, C_{13}, C_{23}, C_0)$ where

$$C_{12} = \text{inf spectrum}\left[\frac{1}{2\mu}(P_{12}^2 + P_{13}^2) + \frac{1}{M}\vec{P}_{12}\cdot\vec{P}_{13} - \frac{2e^2}{|r_{12}|}\right]$$

$$C_{13} = \text{inf spectrum}\left[\frac{1}{2\mu}(P_{12}^2 + P_{13}^2) + \frac{1}{M}\vec{P}_{12}\,\vec{P}_{13} - \frac{2e^2}{|r_{13}|}\right]$$

$$C_{23} = \text{inf spectrum}\left[\frac{1}{2\mu}(P_{12}^2 + P_{13}^2) + \frac{1}{M}\vec{P}_{12}\cdot\vec{P}_{13} + \frac{e^2}{|r_{12} - r_{13}|}\right]$$

$$C_0 = \text{inf spectrum}\left[\frac{1}{2\mu}(P_{12}^2 + P_{13}^2) + \frac{1}{M}\vec{P}_{12}\cdot\vec{P}_{13}\right]$$

It is not apparent that C is in fact the He^+ binding energy, for while Hunziker's theorem eliminates the intercluster potentials, *it does not eliminate intercluster Hughes-Eckart terms*. However, this is just an apparent problem. For $H_{C.M.} = P^2/2m{+}M$ depends on coordinates independent of the Hamiltonians, H_α, whose inf spectrum $H_\alpha = C_\alpha$. Thus

$$\text{inf spectrum } (H_\alpha + H_{C.M.}) = \text{inf } H_\alpha + \text{inf } H_{C.M.} = \text{inf } H_\alpha = C_\alpha.$$

Thus, for example:

$$C_{12} = \text{inf spectrum}\left[H_{C.M.} + \frac{1}{2\mu}(P_{12}^2 + P_{13}^2) + \frac{1}{M}\vec{P}_{12}\cdot\vec{P}_{13} - \frac{2e^2}{|r_{12}|}\right]$$

$$= \text{inf spectrum}\left[\frac{P_1^2}{2M} + \frac{P_2^2}{2m} + \frac{P_3^2}{2m} - \frac{2e^2}{|r_{12}|}\right]$$

$$= \text{inf spectrum}\left[\frac{P_1^2}{2M} + \frac{P_2^2}{2m} - \frac{2e^2}{|r_{12}|}\right] + \text{inf spectrum}\left[\frac{P_3^2}{2m}\right]$$

$$= \text{inf spectrum}\left[\frac{P_{12}^2}{2\mu} - \frac{2e^2}{|r_{12}|}\right] = He^+ \text{ binding energy.}$$

In general, we can use the fact that

$$\inf(\text{spec } \tilde{H}_D) = \inf \text{spec } (\tilde{H}_D + H_{C.M.}) = \inf \text{spec } H_D$$

and conclude:

PROPOSITION VII.6. *The* C *of Theorems VII.4 and VII.5 is* $\min_{D \in \mathcal{C}_2} \inf \sigma(H_D)$ *where* $\mathcal{C}_2$ *is the decompositions into two clusters and* H_D *is the Hamiltonian without the center of mass removed.*

Proof. We need only show that for any n-component cluster $D = \{C_1, \ldots, C_n\}$ there is a two-component cluster $D' = \{C_1', C_2'\}$ with $\inf \sigma(H_{D'}) \leq \inf \sigma(H_D)$, because of our discussion about the center of mass above. But taking $C_2' = C_n$; $C_1' = \bigcup_{i=1}^{n-1} C_i$, we get the required inequality, in fact, if $n > 2$, $\inf \sigma_{cont}(H_{D'}) \leq \inf \sigma(H_D)$.

Appendix 2 to VII.3. Bound States of Helium

To determine the dimension of $\mathcal{H}_{P.P.}$ for the Helium atom is an interesting problem both physically and for its testimony to the power of abstract analysis to give qualitative information about operators which are not exactly soluable.

In 1951, in a remarkable paper [139], Kato demonstrated that the Helium atom Hamiltonian, with the physical masses had at least 25,585 bound states.[11] Several years later Žislin demonstrated that any "atom" or "positive ion" has infinitely many bound states [140].[12] The reason

[11] This counted angular degeneralies and represented 42 "principle" shells or 903 distinct states not counting states with the same $\vec{J}$ and E but different J_z as distinct.

[12] Žislin and Sigalov [141] showed this holds true even if electron statistics was included.

Kato didn't obtain this result was that he was unable to get an ideal estimate on the bottom of the continuum for H_{He} – basically because of the Hughes-Eckart terms.[13] Since Hunziker's theorem tells us exactly where the continuum occurs we should be able to improve Kato's result directly; in fact this is correct:

THEOREM VII.7. $\tilde{H}_{He}$ *has infinitely many bound states.*

Proof. By Hunziker's theorem, $\inf \sigma_{cont}(\tilde{H}_{He}) = \inf \sigma(H_{He+})$. Following Kato, for any N, we find an N-dimensional space V_N and an ε so that $\langle\Psi|\tilde{H}_{He}|\Psi\rangle \le [\inf \sigma(H_{He+}) - \varepsilon]\langle\Psi|\Psi\rangle$ for $\Psi \in V_N$. Thus, by the min-max principle, $\tilde{H}_{He}$ has at least N bound states (for any N). Write

$$\tilde{H}_{He} = \frac{P_1^{\,2}}{2\mu} + \frac{P_2^{\,2}}{2\mu} + \frac{\vec{P}_1 \cdot \vec{P}_2}{M} - \frac{2e^2}{r_1} - \frac{2e^2}{r_2} + \frac{e}{r_{12}}.$$

Let $E_0 = (-\frac{1}{2}\mu c^2)(\frac{2e^2}{hc})^2 = \inf \sigma(He^+)$, and let $\Psi(\vec{r}_1, \vec{r}_2) = \Psi_1(\vec{r}_1)\,\tilde{\Psi}(\vec{r}_2)$ where $\Psi_1(r_1)$ is the eigenfunction of $\frac{P_1^{\,2}}{2\mu} - \frac{2e^2}{r_1}$ and $\tilde{\Psi}(\vec{r}_2)$ is in some N-dimensional subspace spanned by n bound states of $\frac{P_2^{\,2}}{2\mu} - \frac{e^2}{r_2}$,[14] so that

$$\langle\tilde{\Psi}|\frac{P^2}{2_\mu} - \frac{e^2}{r_2}|\tilde{\Psi}\rangle \le -\varepsilon_N \langle\tilde{\Psi}, \tilde{\Psi}\rangle \text{ for some fixed positive } \varepsilon_N.$$

Then:

$$\langle\Psi|\frac{P_1^{\,2}}{2\mu} - \frac{2e^2}{r_1}|\Psi\rangle = \langle\tilde{\Psi}|\tilde{\Psi}\rangle\langle\Psi_{1s}|\frac{P_1^{\,2}}{2\mu} - \frac{2e^2}{r_1}|\Psi_{1s}\rangle = E_0\langle\Psi, \Psi\rangle$$

and

---

[13] When the mass of the "nucleus" was infinite so no Hughes-Eckert terms appeared, Kato showed infinitely many bound states exist.

[14] Note the absence of the $2e^2$. This is because, we will use one nuclear charge to cancel the interelectron repulsion; i.e., we will use the fact that the first electron cannot shield more than one unit of nuclear charge.

$$\langle\Psi|\vec{P}_1\cdot\vec{P}_2|\Psi\rangle = \langle\Psi_{1s}|\vec{P}_1|\Psi_{1s}\rangle\cdot\langle\tilde{\Psi}|\vec{P}_2|\tilde{\Psi}\rangle = 0$$

and

$$\langle\Psi|\frac{P_2^{\,2}}{2\mu}-\frac{e^2}{r_2}|\Psi\rangle = \langle\Psi_{1s}|\Psi_{1s}\rangle\,\langle\tilde{\Psi}|\frac{P_2^{\,2}}{2\mu}-\frac{e^2}{r_2}|\tilde{\Psi}\rangle \le -\varepsilon_N\langle\Psi,\Psi\rangle.$$

Finally,

$$\langle\Psi|\frac{1}{r_{12}}-\frac{1}{r_2}|\Psi\rangle = \langle\tilde{\Psi}|[\textstyle\int d^3r_1\frac{1}{r_{12}}|\Psi_{1s}(r_1)|^2]-\frac{1}{r_2}|\tilde{\Psi}\rangle = \langle\Psi_2|\frac{f(r_2)}{r_2}|\Psi_2\rangle \le 0$$

where

$$f(r_2) = 1 - \int_{r_1\le r_2} d^3r_1|\Psi_{1s}(r_1)|^2 \ge 0.$$

Thus

$$\langle\Psi|H_{He}|\Psi\rangle \le (E_o-\varepsilon_N)\langle\Psi|\Psi\rangle.$$

This completes the proof. ∎

Remarks

1. This proof which is a trivial combining of Kato's paper [139] with Hunziker's theorem is technically an enormous simplification of Žislin's procedure.

2. This proof clearly extends to arbitrary atoms without Fermi-statistics. To prove the theorem for the atomic case would require extending Hunziker's theorem to include statistics in some way. This shouldn't be hard and warrants further study.

3. This question has since been treated in a more directly physical manner by Simon [143]. In addition, it was proven there that Hunziker's theorem does hold with statistics.

Appendix 3 to VII.3. Operators with Hilbert-Schmidt Fibers

In this appendix, we should like to discuss the details of the proof of I in Theorem VII.5 in more detail by putting the notion of p-space δ-functions in kernels into a more mathematical guise. To say an operator from $L^2(\vec{r}_1,\ldots,\vec{r}_N)$ to itself has a $\delta(\vec{P}_i - \vec{P}_i)$ in its kernel says merely that it commutes with translations $\vec{r}_j \to \vec{r}_j + \delta_{ij}\vec{a}$, or equivalently that it commutes with multiplication by $f(\vec{P}_i)$ when we "transform" the operator to momentum space, i.e., consider $\mathcal{F}T\mathcal{F}^{-1}$ where $\mathcal{F}$ is the Fourier transform. Thus we wish to consider operators on $L^2(\vec{P}_1,\ldots,\vec{P}_N)$ = commuting with $\vec{P}_1,\ldots,\vec{P}_r$. This can be done most easily by realizing $\mathcal{H}$ as L^2 functions on $R^{3r}(=(\vec{P}_1,\ldots,\vec{P}_r))$ with values in $L^2(P_{r+1},\ldots,P_N) = M$. Thus we are in the position of looking at $\mathcal{H}$ as $L^2(\Omega; M)$, the functions from Ω to M with $\int d\mu \|f(\vec{x})\|^2 < \infty$.[15] In this guise, an operator commuting with $\vec{P}_1,\ldots,\vec{P}_r$ is merely an operator on $L^2(\Omega; M)$ commuting with a functions in $L^\infty(\Omega)$. It is quite easy to prove:

PROPOSITION VII.8. If $T : L^2(\Omega; M) \to L^2(\Omega; M)$ so that $T(gf) = g(Tf)$ for all $g \in L^\infty(\Omega)$, all $f \in L^2(\Omega)$, then T has the form

$$(Tf)(x) = T(x)f(x)$$

where $T(x)$ is a map from $M \to M$. Moreover,

$$\operatorname*{ess\,sup.}_x \|T(x)\|_M = \|T\|_{L^2(\Omega;\, M)}. \blacksquare$$

DEFINITION. If T is an operator obeying the conditions of VII.8, we say T is fibered and call $T(x)$ the fibers of T. If almost every $T(x)$ is Hilbert-Schmidt on M and ess sup. $\|T(x)\|_{H.S.} < \infty$, we say T is an operator with Hilbert-Schmidt fibers. We write $\|T\|_{F.H.S.} = \operatorname*{ess\,sup.}_x \|T(x)\|_{H.S.}$.

It then follows directly from the formula $\|AB\|_{H.S.} \le \|A\|_{H.S.}\|B\|$, that the following proposition holds:

15 Which is nothing but a trivial direct integral $\int_\Omega^\oplus d\mu\, M$.

PROPOSITION VII.9. Let $T,S : L^2(\Omega; M) \to L^2(\Omega; M)$ be fibered and let T have Hilbert-Schmidt fibers and:

$$\|TS\|_{F.H.S.} \leq \|T\|_{F.H.S.} \|S\|_{L^2(\Omega;\, M)}. \blacksquare$$

We can now explain what we were doing in Part I of the proof of Theorem VII.5. We started out with

$$L^2(\vec{P}_2{}',\ldots,\vec{P}_{N-1}{}',\vec{P}_N{}') = L^2(\vec{P}_2{}',\ldots,\vec{P}_{N-1}{}';\, L^2(\vec{P}_N{}')).$$

With respect to this fibering both $(G_o^{1/2}\, V_{i_N j_N}\, G_o^{1/2})$ and $R_{D_{N-1}}$ are fibered and the first has Hilbert-Schmidt fibers so their product has Hilbert-Schmidt fibers. Next we viewed

$$L^2(\vec{P}_2{}',\ldots,\vec{P}_{N-1}{}',\vec{P}_N{}') \text{ as } L^2(\vec{P}_2{}',\ldots,\vec{P}_{N-2}{}';\, L^2(\vec{P}_{N-1}{}',\vec{P}_N{}')).$$

$(G_o^{1/2}\, V_{i_N j_N}\, G_o^{1/2})\, R_{D_{N-1}}$ is fibered with respect to this fibering. While its fibers are not Hilbert-Schmidt, if we regard

$$L^2(\vec{P}_{N-1}{}',\vec{P}_N{}') = L^2(\vec{P}_{N-1}{}') \otimes L^2(\vec{P}_N{}') = L^2(\vec{P}_{N-1}{}';\, L^2(\vec{P}_N{}'))$$

the fibers of the fibers (which are the fibers with respect to the first fibering!) are Hilbert-Schmidt by Proposition VII.9. $G_o^{1/2}\, V_{i_{N-1} j_{N-1}}\, G_o^{1/2}$ has Hilbert-Schmidt fibers with respect to $L^2(\vec{P}_{N-1}{}',\vec{P}_N{}') = L^2(\vec{P}_N{}';\, L^2(\vec{P}_{N-1}{}'))$ so that $(G_o^{1/2}\, V_{i_N j_N}\, G_o^{1/2}) R_{D_{N-1}} (G_o^{1/2} V_{i_{N-1} j_{N-1}}\, G_o^{1/2})$ has Hilbert-Schmidt fibers in $L^2(\vec{P}_{N-1}{}',\vec{P}_N{}')$ over $L^2(\vec{P}_2{}',\ldots,\vec{P}_{N-2}{}')$. Since $R_{D_{N-2}}$ is fibered relative to this fibering,

$$(G_o^{1/2}\, V_{i_N j_N}\, G_o^{1/2}) R_{D_{N-1}} (G_o^{1/2}\, V_{i_{N-1} j_{N-1}}\, G_o^{1/2}) R_{D_{N-2}}$$

has Hilbert-Schmidt fibers by VII.9. We go on playing this game until we are done.

Appendix 4 to VII.3. An Alternate Proof that I_S *is Compact*

We sketch here an alternate proof that I_S is compact which avoids the necessity for induction. The proof would work just as well for I when $V \in L^2$ and so provides an alternate proof of the condition used by Hunziker. We use crucially the following lemma:

LEMMA VII.10. *Let* $I(z)$ *be an operator-valued analytic function in a connected domain* D. *If* $I(z_n)$ *is compact for a sequence* z_n *with a finite limit point the* $I(z)$ *is compact for all* $z \in D$.

Proof. Suppose $I(w)$ is not compact. Since the compact operators are closed, by the Hahn-Banach theorem we can find $\ell \in B(\mathcal{H})^*$ with $\ell(I(w)) = 1$ and $\ell \equiv 0$ on the compacts. Since $\ell(I(z))$ is analytic and $\ell(I(z_n)) = 0$, this is impossible. ∎

Note. An essentially identical theorem is proven by Hunziker in an appendix of [145] but by a different method.

To prove $I_S(E)$ is compact, we need only do it for E real and very negative. For such E, the diagram expansion is $\| \;\|$-convergent. Since the compact operators are $\| \;\|$-closed, we need only prove each connected diagram is compact. For $V_{ij} \in R$, a direct computation shows these are Hilbert-Schmidt (the kernels are explicitly known in p-space). By a limiting argument, the diagrams for $V_{ij} \in R + (L^\infty)_\varepsilon$ are compact also.

APPENDIX

SOME MATHEMATICAL BACKGROUND

We collect here background theorems on the structure of operators in Hilbert space. Coherent general references on Hilbert space include Halmos [T12], Chap. 1; Kato [T17], Chaps. III, V; Lorch [T20], Chaps. 1-3; Nelson [T23] and Riesz-St. Nagy [T25].

(a) *Basic Definitions for Self-Adjointness*

For pro forma purposes, let us define Hilbert space:

DEFINITION A.1. A sesquilinear form on a complex vector space $\mathcal{H}$ is a map $B : \mathcal{H} \times \mathcal{H} \to \mathbb{C}$ so that $B(x_0, \cdot)$ is linear and $B(\cdot, y_0)$ is conjugate linear.

DEFINITION A.2. A complex vector space endowed with a sesquilinear form $< , >$ obeying $<x, x> > 0$ when $x \neq 0$, is called an *inner product space*. If $\mathcal{H}$ is a complete metric space when given the metric $\rho(x,y) = \|x-y\|$ where $\|x\| = \sqrt{<x, x>}$, then $\mathcal{H}$ is a *Hilbert space*.

Henceforth $\mathcal{H}$ will stand for a Hilbert space.

THEOREM A.1 (Riesz lemma). *Every continuous linear functional,* f, *on* $\mathcal{H}$ *is of the form:*

$$f(\phi) = <\psi,\phi>$$

for some ψ. *Every bounded sesquilinear form (i.e.,* B *obeying* $|B(\psi,\phi)| \leq C\|\psi\| \, \|\phi\|$) *is of the form* $B(\psi,\phi) = <\psi,A\phi>$ *with* A *a uniquely determined bounded operator.*

Proof. See Halmos [T12], pp. 31-32.

An immediate application of Theorem A.1 is to define the adjoint of a *bounded* operator:

If A is bounded $B(\psi,\phi) = \langle A\psi,\phi\rangle$ obeys $|B(\psi,\phi)| \le \|A\|\,\|\psi\|\,\|\phi\|$ so we can:

DEFINITION A.3. Given a bounded operator, A, we define A^*, the adjoint of A, by the identity

$$\langle\psi, A^*\phi\rangle = \langle A\psi,\phi\rangle$$

* * *

One often has occasion to use various types of convergence other than the usual norm convergence. In particular, 5 topologies are needed, 2 on vectors and 3 on operators:

DEFINITION A.4. We say ψ_n *converges to* ψ *in norm* if and only if $\|\psi_n-\psi\| \to 0$. We will write $\psi_n \xrightarrow{\|\,\|} \psi$ or $\|\ \|\text{-}\lim \psi_n = \psi$.

DEFINITION A.5. We say ψ_n *converges to* ψ *weakly* if and only if $\langle\phi,\psi_n-\psi\rangle \to 0$ for each fixed ϕ. We will write $\psi_n \overset{w}{\to} \psi$ or $\text{w-}\lim \psi_n = \psi$.

Remarks

1. $\psi_n \xrightarrow{\|\,\|} \psi$ implies $\psi_n \overset{w}{\to} \psi$.

2. The converse of 1 is false, for let ψ_n be an orthonormal basis. Since $\Sigma|\langle\phi,\psi_n\rangle|^2 < \infty$, $\langle\phi,\psi_n\rangle \to 0$, i.e., $\psi_n \overset{w}{\to} 0$ but $\|\psi_n\| = 1$ so $\psi_n \not\xrightarrow{\|\,\|} 0$.

DEFINITION A.6. Let A_n be a sequence of bounded operators. We say A_n *converges in norm* to A if and only if $\|A_n - A\| \to 0$. We write $A_n \xrightarrow{\|\,\|} A$ or $\|\ \|\text{-}\lim A_n = A$.

DEFINITION A.7. Let A_n be a sequence of bounded operators. We say A_n *converges strongly* to A if and only if $A_n\psi \xrightarrow{\|\,\|} A\psi$ for all $\psi \in \mathcal{H}$. We will write $A_n \overset{s}{\to} A$ or $\text{s-}\lim A_n = A$.

DEFINITION A.8. Let A_n be a sequence of bounded operators. We say A_n *converges weakly* to A if and only if $A_n\psi \xrightarrow{W} A\psi$ for all $\psi \in \mathcal{H}$ (equivalently if and only if $\langle\phi, A_n\psi\rangle \to \langle\phi, \psi\rangle$ for all $\phi, \psi \in \mathcal{H}$). We will write $A_n \xrightarrow{W} A$ or w-lim $A_n = A$.

Remarks

1. $A_n \xrightarrow{\| \|} A$ implies $A_n \xrightarrow{S} A$ implies $A_n \xrightarrow{W} A$.

2. Let $|\psi\rangle\langle\phi|$ be the operator $(|\psi\rangle\langle\phi|)\eta = \langle\phi,\eta\rangle\psi$. Let ϕ_n be an orthonormal basis and let $A_n = |\phi_n\rangle\langle\phi_1|$; $B_n = A_n^* = |\phi_1\rangle\langle\phi_n|$. Then $A_n \xrightarrow{W} 0$ but $A_n \not\xrightarrow{S} 0$; $B_n \xrightarrow{S} 0$ but $B_n \not\xrightarrow{\| \|} 0$.

3. $A \rightsquigarrow A^*$ is continuous in $\| \ \|$ and in the weak operator topology.

4. $A \rightsquigarrow A^*$ is *not* continuous in the strong operator topology. In the example of 2, $B_n \xrightarrow{S} 0$ but $B_n^* (= A_n) \not\xrightarrow{S} 0$.

* * *

Unfortunately not all physical operators are bounded; for bounded operators have bounded expectation values and the energy, momentum, angular momentum, and position do not have bounded expectation values. Unbounded operators cannot be defined everywhere (at least if they are hermitean) for:

THEOREM A.2 (Hellinger-Toeplitz). *An everywhere defined operator* A, *obeying* $\langle\phi, A\psi\rangle = \langle A\phi, \psi\rangle$ *is bounded.*

Proof. This is a simple consequence of either the closed graph theorem (Lorch [T20], p. 44) or the Banach-Steinhaus principle (Lorch [T20], pp. 45-46). ∎

The domain of definition of an operator A will be denoted D(A). Although for a symmetric operator Theorem A.2 tells us that D(A) cannot be all of Hilbert space, there is an important class of operators for which D(A) is dense. For example, when A = X, the position operator on $\mathbb{R}$, and $\psi = (1 + x^2)^{-1/2}$, $x\psi(x) \notin L^2(\mathbb{R})$ so $x\psi$ doesn't make sense *as an element of* $\mathcal{H}$; thus $\psi \notin D(X)$. One usually defines $D(X)$ by

$$D(X) = \{\psi \in L^2 | \int x^2 \, |\psi(x)|^2 \, dx < \infty\}$$

and then $D(X)$ is dense.

Unbounded operators also present problems in the definition of adjoints. We cannot quite use the method we used for bounded operators. However, a simple modification suffices:

DEFINITION A.9. Let A on D(A) be given. We say $\psi \in \tilde{D}$ if and only if the map $\phi \rightsquigarrow \langle\psi, A\phi\rangle$ defined for $\phi \in D(A)$ is continuous and so extendable to all $\phi \in \mathcal{H}$. The *adjoint*, A^*, is defined with $D(A^*) = \tilde{D}$ so that $\langle A^*\psi, \phi\rangle = \langle\psi, A\phi\rangle$.

It is not hard to see that A^* is a linear operator. For Hermitean operators, we already know much about adjoints:

DEFINITION A.10. We say A is *Hermitean* if and only if $\langle\phi, A\psi\rangle = \langle A\phi, \psi\rangle$ all $\phi, \psi \in D(A)$.

THEOREM A.3. *If* A *is Hermitean,* A^* *is an extension of* A (*which we write* $A \subset A^*$), *i.e.,* $D(A) \subset D(A^*)$ *and* $A^*\phi = A\phi$ *for all* $\phi \in D(A)$. ∎

That $A \neq A^*$ is possible for Hermitean operators leads to the most glorious subtlety of mathematical physics: the distinction between self-adjoint and Hermitean operators.

DEFINITION A.11. We say A is *self-adjoint* if $A = A^*$.

That this distinction is not mere mathematical nit-picking is seen most readily by considering a free particle in a one-dimensional box. As a first guess, we might take D(H) to be those functions, f, vanishing *near* the endpoints of the box for which $-\dfrac{d^2}{dx^2}$ makes sense. Such an operator is Hermitean but not self adjoint. It has a four (real) parameter family of self-adjoint extensions (essentially parameterized by U(2)), which are characterized by *boundary conditions* (e.g., periodic, reflecting, etc.).

Thus, self-adjointness is in some sense Hermiticity plus "sufficiently-many" boundary conditions. We will see in (c), that the Hamiltonian of a physical system must be self-adjoint and not merely Hermitean (Note: For a fuller discussion of the above example, see Wightman [122]).

For an operator to be self-adjoint, we must choose precisely the correct domain, but this is often hard to do and we expect some "inessential" features of the domain to be unimportant. For example, when studying the operator $H_o = -\nabla^2$ on R^3, we expect no boundary conditions at infinity. We thus would think that giving all C^∞ functions of compact support as domain would somehow characterize H_o sufficiently. That it does is made precise by the notion of "closure" and "essentially self-adjoint."

DEFINITION A.12. An operator A with domain D is called *closed* if and only if $\Gamma(A) = \{<\phi,\psi> \in \mathcal{H} \times \mathcal{H} | \phi \in D(A); \psi = A\phi\}$ (called the graph of A) is closed; equivalently, A is closed if $\phi_n \in D(A)$; $\phi_n \xrightarrow{\|\cdot\|} \phi$, $A\phi_n \xrightarrow{\|\cdot\|} \psi$ imply $\phi \in D(A)$ and $\psi = A\phi$.

DEFINITION A.13. An operator is called closable if $\overline{\Gamma(A)}$ is the graph of an operator which we call $\bar{A}$. $\bar{A}$ is the smallest closed extension of A.

THEOREM A.4. *If* A^* *is densely defined, then* A *is closable and* $\bar{A} = (A^*)^*$. In particular, any Hermitean operator is closable.

THEOREM A.5. *The following are equivalent for a Hermitean operator* A.

(a) $\bar{A}$ *is self-adjoint.*

(b) A *has a unique self-adjoint extension.*

(c) $A^* = A^{**}$.

(d) A^* *is Hermitean.*

If any (and thus all) of the above hold, we say that A is *essentially self-adjoint.* If A is self-adjoint and $\mathfrak{D} \subset D(A)$ is such that $A \restriction \mathfrak{D}$ is essentially self-adjoint, we say $\mathfrak{D}$ is a *core* for A. For proofs of Theorems A.4 and A.5, see Kato [T17], p. 269.

Evidently, one needs general criteria for self-adjointness in order to study general problems. One of the most useful is:

THEOREM A.6. A *is self-adjoint if and only if one (and hence both) of the following holds:*

(a) Ran $(A \pm i) = \mathcal{H}$ *or*

(b) A *is closed and* Ker $(A^* \pm i) = 0$ (*where* Ker $A = \{\Psi \in D(A) | A\Psi = 0\}$ *and* Ran $A = \{A\Psi | \Psi \in D(A)\}$).

A *is essentially self-adjoint if and only if:*

(A)′ Ran $(A \pm i)$ *is dense. Or*

(b)′ Ker $(A^* \pm i) = 0$.

Proof. See Riesz-Nagy [T25], pp. 325-328.

As one of its consequences, this has the following "perturbation" theorem:

THEOREM A.7 (The Kato-Rellich theorem). *Let* A *be self-adjoint and let* B *be a Hermitean operator obeying:*

(a) $D(B) \supset D(A)$.

(b) *There is an* $a < 1$ *and* $b > 0$ *so that*

$$\|B\psi\| \leq a\|\psi\| + b\|\psi\|$$

for all $\psi \in D(A)$.

Then $A + B$ *defined on* $D(A)$ *is self-adjoint.* $\mathfrak{D}$ *is a core for* $A + B$ *if and only if it is a core for* A.

Proof. See Kato [T17], pp. 287-289.

We thus define:

DEFINITION A.14. If A is given and B obeys the condition of Theorem A.7, we say B is *Kato-small* relative to $A(B \underset{\text{T.K.}}{<} A)$. If a can be chosen arbitrarily small (with b, in general, dependent on a) we say A is *Kato-tiny* relative to A. $(B \underset{\text{T.K.}}{<<} A)$.

In Section II.2, we prove a result (due to Kato, Lions, Lax, Milgram and Nelson) which is similar to Theorem A.7 but uses expectation values instead of norms.

(b) *Spectral Theory for Self-Adjoint Operators*

Once basic mathematical objects have been identified, one of the main problems of mathematicians is to find a general "structure" theorem for the object, i.e., a complete canonical description of such objects; prime examples of such theorems are the Gel'fand-Naimark theorem for commutative C^*-algebras and the structure theorem for finite abelian groups. There is a general structure theorem for self-adjoint operators on separable Hilbert spaces; the justly famous spectral theorem. It has many different forms; undoubtedly, the most cogent is:

THEOREM A.8 (The Spectral Theorem). *Let* A *be an operator on a separable Hilbert space,* $\mathcal{H}$. *Then, there is a collection of measures on the real line:* $\mu_1 \ldots, \mu_n$ *(finite or infinite) and a unitary map*

$$U : \to L^2(\cup R, \mu)$$

[*where* $\cup R$ *is a union of real lines, one for each measure;* μ *is the measure* μ_n *on the* n^{th} R; *thus* $L^2(\cup R,\mu) \equiv$ *functions* $\psi(x;n)$; $x \in R$; $n = 1,\ldots$ *with* $\sum_n \int |\psi(x,n)|^2 \, d\mu_n(x) < \infty$], *so that*

(a) $\phi \in D(A)$ *if and only if the function* $\psi = U\phi$ *has the property* $\sum_n \int |\psi(x,n)|^2 x^2 \, d\mu_n(x) < \infty$.

(b) *If* $\psi \in U[D(A)]$, *then* $(UAU^{-1}\psi)(x,n) = x\psi(x;n)$.

Proof. See Nelson [T23], Section I.5.

Remarks

1. To be a true structure theorem, we must pick the μ's canonically in some sense. There is a canonical choice (not of measure, but of measure class; i.e., two measures absolutely continuous with respect to each other are not distinguished) given by the Hahn-Hellinger *multiplicity theory*, see e.g., Nelson [T23], Section I.6.

2. This is a generalization of the more usual finite dimensional spectral theorem. In that case, when A is a Hermitean matrix without degen-

erate eigenvalues, say $\lambda_1,\ldots,\lambda_N$, we pick μ to be the pure point measure $\mu = \sum_n \delta_{\lambda_n}$ i.e., $\mu(A) = \#$ of λ_i in A. Then $L^2(R,\mu) = R^N$; the vector $(a_1,\ldots,a_N)$ corresponds to the function $f(\lambda_i) = a_i$ [$R-\{\lambda_1,\ldots,\lambda_N\}$ has measure 0 so f is "undefined" there]. Thus $X(a_1,\ldots,a_N) = (\lambda_1 a_1,\ldots,\lambda_N a_N)$ so A is represented as a multiplication operator. If A has degenerate eigenvalues more than one μ_i and more than one R will be needed.

3. The spectral theorem is normally stated in the $\int \lambda dE_\lambda$ form, a theorem without the conceptual clarity or simplicity of Theorem A.8. Since the E_i's are useful, both forms are important.

4. In the form given above, the spectral theorem is obviously a rigorous analogue of the familiar Dirac notation.

The "spectral" is not terribly evident in this form of the spectral theorem. Let us first recall the usual definition of spectrum and link it up to the "spectral measures," μ_n of Theorem A.8:

DEFINITION A.15. A complex number z is said to be in the *resolvent set*, $\rho(A)$, for a self-adjoint operator A if and only if $\{\langle (A-z)\phi,\phi\rangle \mid \phi \in D(A)\}$ is the graph of a bounded operator. $\mathbb{C} - \rho(A) \equiv \sigma(A)$, is called the *spectrum* of A.

THEOREM A.9. *If A is self-adjoint, $\sigma(A)$ is a closed subset of R. In particular, if* $\operatorname{Im} z \neq 0$, $(A-z)^{-1}$ *exists and* $\|(A-z)^{-1}\| \leq |\operatorname{Im} z|^{-1}$.

Proof. See e.g., Halmos, pp. 54-55. ∎

The connection between the spectral measures and the spectrum is simple:

THEOREM A.10. *Let z be real. Then $z \in \rho(A)$ if and only if $\exists \delta$ so that $\mu_n((z-\delta, z+\delta)) = 0$ for all spectral measures μ_n.*

Proof. Let such a δ exist. Let B be defined in the spectral representation by $(B\psi)(x;n) = (x-z)^{-1}\psi(x;n)$. Then B is a bounded operator with $\|B\| \leq \delta^{-1}$, and $B = (A - z)^{-1}$. Conversely, if no such δ exists, it is easy to construct $\{\phi_n\}$ with $\|\phi_n\| = 1$ and $\|(A-z)\phi_n\| < 1/n$. Thus $(A-z)^{-1}$ cannot exist as a bounded operator. ∎

This last theorem has a simple rewording in terms of support of a measure. In general, given a measure, μ, there is not a largest set of measure 0 (consider Lebesgue measure) but there is a largest *open* set of measure 0 (because there is a countable base of open sets).

DEFINITION A.16. The *support of a measure* is the complement of the largest open set of measure 0. The *support of a family of measures* is the complement of the largest open set which has measure 0 with respect to all the measures of the family.

Thus, Theorem A.10 is equivalent to:

THEOREM A.10′. *$\sigma(A)$ is the support of the family of spectral measures.*

For many purposes, the spectrum is a coarse invariant. For example, the ordinary position operator and an operator with a complete set of eigenvectors whose eigenvalues are all rationals have the same spectrum, namely $\mathbb{R}$. It is often useful to squeeze more information out of the spectral measures before looking at their supports. We first recall:

DEFINITION A.17. A measure μ is called *pure point* if it can be written as a sum

$$\mu = \Sigma\, a_i \delta_{x_i} \quad \text{where } \delta_{x_i} \text{ is the measure:}$$

$$\delta_{x_i}(\Omega) = \begin{cases} 0 & \text{if } x \notin \Omega \\ 1 & \text{if } x \in \Omega \end{cases}$$

A measure is said to have no pure points if $\mu(\{x\}) = 0$ for all x.

DEFINITION A.18. A measure μ is said to be *absolutely continuous* (with respect to Lebesgue measure) if $\mu(A) = 0$ whenever A has zero Lebesgue measure. Equivalently, by the Radon-Nikodym theorem (see

Halmos [T11], pp. 128-130 or Berberian [T2], pp. 167-169), μ is of the form $\mu(A) = \int_A f(x)dx$ for a function f locally $L^1(R, dx)$.

DEFINITION A.19. We say a measure, μ, is *singular* (with respect to Lebesgue measure) if and only if μ has no pure points and there is a set A of Lebesgue measure 0 with $\mu(R/A) = 0$.

Remark

Our use of singular is not quite standard. What we have called singular is often called "singular continuous" in which case singular measures include pure point measures.

THEOREM A.11. *Any measure, μ, on* R *has a unique decomposition* $\mu = \mu_{p.p.} + \mu_{a.c.} + \mu_{sing}$ *where:*

$\mu_{p.p.}$ *is a pure point measure,*

$\mu_{a.c.}$ *is absolutely continuous;*

μ_{sing} *is singular.*

Proof. See Berberian [T2], pp. 158-159. ∎

Corresponding to this decomposition theorem, there is a splitting of $\sigma(A)$ and of $\mathcal{H}$. For simplicity, we suppose A is multiplicity free, i.e., only one μ_n is needed. Let $\lambda_1, \ldots, \lambda_n, \ldots$ (finite or infinite) be the pure points of μ. These are precisely the eigenvalues of A. Let $\mathcal{H}_{p.p.}$ be the subspace of $\mathcal{H}$ generated by all the eigenvectors and let $\mathcal{H}_{cont} = \mathcal{H}^{\perp}_{p.p.}$. The spectral measure of $A \upharpoonright \mathcal{H}_{cont}$ is $\mu_{a.c.} + \mu_{sing}$. Let Ω be chosen so $\mu_{a.c.}(\Omega) = 0$, $\mu_{sing}(R/\Omega) = 0$. Let $(P\psi)(x) = \chi_{\Omega}(x)\,\psi(x)$ in the spectral representation for A. Then A and P commute and $A \upharpoonright P\mathcal{H}_{cont}$ has μ_{sing} as its spectral measure and $A \upharpoonright (1-P)\mathcal{H}_{cont}$ has $\mu_{a.c.}$ as its spectral measure. Thus:

THEOREM A.12. *Let* A *be a self-adjoint operator. Then there is a unique decomposition:*

$$\mathcal{H} = \mathcal{H}_{p.p.} \oplus \mathcal{H}_{a.c.} \oplus \mathcal{H}_{sing}$$

so that A *leaves each* $\mathcal{H}_{\dots}$ *invariant and:*

$A \restriction \mathcal{H}_{p.p.}$ *has a complete set of eigenvectors.*

$A \restriction \mathcal{H}_{a.c.}$ *has absolutely continuous spectral measures.*

$A \restriction \mathcal{H}_{sing}$ *has singular spectral measures.*

If $\sigma_{p.p.}$ *is the set of eigenvalues, then* $\sigma(A \restriction \mathcal{H}_{p.p.}) = \overline{\sigma}_{p.p.}$. *We define* $\sigma_{a.c.(sing)}(A) \equiv \sigma(A \restriction \mathcal{H}_{a.c.(sing)})$ *and* $\sigma_{cont}(A) = \sigma(A \restriction \mathcal{H}_{cont}) = \sigma_{a.c.}(A) \cup \sigma_{sing}(A)$ *where* $\mathcal{H}_{cont} = \mathcal{H}_{a.c.} \oplus \mathcal{H}_{sing}$. ∎

Remarks

1. $\sigma_{a.c.}$, σ_{sing} and $\sigma_{p.p.}$ are not necessarily disjoint, nor is $\sigma_{cont} \cup \sigma_{p.p.} = \sigma$ (although $\sigma_{cont} \cup \sigma_{p.p.} = \overline{\sigma}$).

2. Since a singular measure $\mu \neq 0$ has no pure points, its support must be uncountable. Thus, if σ_{sing} is countable, $\sigma_{sing} = \phi$ and $\mathcal{H}_{sing} = 0$.

DEFINITION A.20. Given a self-adjoint, we define $E_{a.c.}$, E_{sing}, $E_{p.p.}$ *as the projections onto* $\mathcal{H}_{a.c.}$, $\mathcal{H}_{sing}$ *and* $\mathcal{H}_{p.p.}$ *respectively.*

* * * *

One crucial thing which the spectral theorem allows us to do is to form arbitrary (complex-valued) bounded measurable functions of a self-adjoint operator A. Given any such function f, we define f(A) by going to a spectral representation and setting.

$$(f(A)\psi)(x;n) = f(x)\psi(x;n).$$

(*Note:* We really mean $Uf(A)U^{-1}$ is defined this way where U is given by Theorem A.8.) This functional calculus is independent of the choice of spectral measure (which is not unique!) and has many nice properties:

THEOREM A.14. *Let* $\phi(f) = f(A)$ *where* f *is a bounded measurable function.*

Then

(a) ϕ *is an algebraic *-homomorphism, i.e.,*

$$\phi(\lambda f + g) = \lambda\phi(f) + \phi(g);\ \phi(\bar{f}) = \phi(f)^*;\ \phi(fg) = \phi(f)\,\phi(g).$$

(b) $\|\phi(f)\| \leq \|f\|_\infty \equiv \sup_x |f(x)|$.

(*Note:* We get equality $\|\phi(f)\| = \|f\|_{(\infty)}$ if $\|\ \|_{(\infty)} \equiv$ the L^∞ norm relative to the spectral measures.)

(c) If $\{\|f_n\|_\infty\}$ is bounded and $f_n(x) \to f(x)$ pointwise for all x, then $\phi(f_n) \xrightarrow{S} \phi(f)$.

(d) If $\psi \in D(A)$, $\phi(f_{id,n})\,\psi \to A\psi$ where $f_{id,n}(x) = \begin{cases} x & |x| < n \\ 0 & |x| \leq n \end{cases}$

(e) $\phi(f)\psi = f(\lambda)\psi$ if $A\psi = \lambda\psi$

(f) If $f(x) \geq 0$ all x, then $\phi(f) \geq 0$, i.e., $\langle\psi,\phi(f)\psi\rangle \geq 0$ all ψ.

Proof. Straight-forward [(c) uses the Lebesgue dominated-convergence theorem].

Remarks

1. Proving this theorem from Theorem A.8 is the reverse of the logical order. The most coherent way of proving A.8 is to first prove A.14 for continuous functions and then use this functional calculus and the Riesz-Markov theorem to construct the spectral measures.

2. If f is a function which is bounded on $\sigma(A)$, we can define f(A) as if f were bounded; f(A) will be a bounded operator.

There are at least three important examples of the functional calculus:

1. If $\lambda \notin \sigma(A)$, $f(x) = (x-\lambda)^{-1}$ is bounded on $\sigma(A)$. f(A) is $(A-\lambda)^{-1}$, the resolvent which we have already discussed.

2. If $f(x) = e^{ixt}$, we get operators $U_A(t) = e^{iAt}$. These operators are unitary and obey $U(t+s) = U(t)\,U(s)$. We return to them in (c) of this appendix.

3. For any Borel set Ω, let $f(x) = \chi_\Omega(x)$ the characteristic function of Ω. Let $E_\Omega = \chi_\Omega(A)$. Since χ is real and $\chi^2 = \chi$, E_Ω is a projection.

DEFINITION A.21. The projections E_Ω are called the *spectral projections* for A.

DEFINITION A.22. A family of projections P_Ω defined for all measurable $\Omega \subset R$ is called a *projection valued measure* (p.v.m.) if:

(a) $\Omega_1,\ldots,\Omega_n,\ldots$ disjoint and $\Omega = \cup\Omega_n$ implies

$$P_\Omega = \text{s-lim} \sum_{n=1}^{N} P_{\Omega_n}$$

(b) $P_R = 1$

If P_Ω is a p.v.m. and $\phi,\psi \in \mathcal{H}$, then $\mu_{\phi,\psi}(\Omega) = \langle\phi,P_\Omega\psi\rangle$ defines a complex measure whose differential we denote by $d\langle\phi,P_\lambda\psi\rangle$. This returns us to the "classical" form of the spectral theorem:

THEOREM A.15. *For any self-adjoint operator,* A, *there is a unique* p.v.m. E_Ω *so that*

$$A = \int \lambda \; dE_\lambda$$

which means $\phi \in D(A)$ *if and only if*

$$\int |\lambda|^2 \; \langle\phi,E_\lambda\phi\rangle < \infty$$

and for all ϕ $\psi \in D(A)$

$$\langle\phi,A\psi\rangle = \int \lambda \; d\langle\phi,E_\lambda\psi\rangle.$$

Remarks

1. This form of the spectral theorem says that there is a 1–1 correspondence between p.v.m's and self-adjoint operators.

2. Mackey [T21], p. 76 has emphasized that it is through this theorem that self-adjoint operators enter physics. In Mackey's formulation, observables are p.v.m's; for given a vector state, ϕ, and an observable, $\mathcal{O}$, there is a measure $\mu_\phi^{(\mathcal{O})}$ with $\mu_\phi^{(\mathcal{O})}(R) = 1$ and with $\mu_\phi^{(\mathcal{O})}(\Omega) =$ the

probability that the value of $\mathcal{O}$ in state ϕ will lie in Ω. $P_\Omega^{(\mathcal{O})}$ is defined so that $\langle\phi, P_\Omega \phi\rangle = \mu_\phi^{(\mathcal{O})}(\Omega)$, and $\mathrm{Exp}_{(\phi)}^{\mathcal{O}} = \int \lambda \langle\phi, P_\lambda \phi\rangle$. By the spectral theorem then, each observable has an associated self-adjoint operator A with $\mathrm{Exp}_\phi(\mathcal{O}) = \langle\phi, A\phi\rangle$ and conversely, every self-adjoint operator defines an observable in the Mackey sense.

Let use denote the spectral projection for A as $E_\Omega(A)$. There is some reason for studying the dependence of $E_\Omega(A)$ on A. At least two theorems are of some direct interest:

THEOREM A.16. *Let* $a < b$ *with* $a, b \notin$ spec. (A). *Let* $(A_n - E)^{-1} \xrightarrow{\|\ \|} (A-E$ *for some* E. *Then for* n *sufficiently large,* $a, b \notin$ spec (A_n) *and* $E_{[a,b]} \xrightarrow{\|\ \|} E_{[a,b]}(A)$.

Proof. See Riesz-Nagy [T25], p. 372. ∎

The idea is to write $E_{[a,b]}(B) = (2\pi i)^{-1} \oint_C d\lambda\, (B-\lambda)^{-1}$ where C is c curve encircling [a,b].

THEOREM A.17. *Let* A, A_n *be self-adjoint and let* a *not be an eigenvalue of* A. *Let* $(A_n - E)^{-1} \xrightarrow{\|\ \|} (A-E)^{-1}$ *for some* E. *Then*

$$E_{[a,\infty)}(A_n) \xrightarrow{S} E_{[a,\infty)}(A).$$

Proof. See Kato [T17], p. 432 where a stronger theorem is proved. ∎

As these theorems suggest, the spectral projection are related to the spectrum in a simple way. They are also related to the notion of discrete spectrum:

DEFINITION A.23 $z \in \sigma(A)$ [A self-adjoint] is said to be a *discrete point* if:

(a) z is an isolated point of $\sigma(A)$, i.e., for some δ,

$$\sigma(A) \cap \{w \mid 0 < |z-w| < \delta\} = \phi.$$

(b) As an eigenvalue, z has finite multiplicity.

THEOREM A.18.

(a) $z \in \sigma(A) \iff (\forall\delta)\ (E_{(z-\delta,z+\delta)}(A) \neq 0)$.

(b) $z \in \sigma_{disc}(A) \iff (\forall\delta)\ (E_{(z-\delta,z+\delta)} = 0)$ *and* $(\exists\delta)\ (E_{(z-\delta,z+\delta)})$ *is finite dimensional.*

Proof. Straight-forward. ∎

DEFINITION A.24. $\sigma_{ess}(A) = \sigma(A)/\sigma_{disc}(A)$. The points of σ_{ess} are called *the essential spectrum.*

THEOREM A.19. $z \in \sigma_{ess}(A)$ *if and only if one or more of the following holds:*

(a) $z \in \sigma_{cont}(A)$.

(b) z *is a limit point of* $\sigma_{p.p.}(A)$, *i.e.*, $z \in \overline{[\sigma_{p.p.}(A) - \{z\}]}$)

(c) $z \in \sigma_{p.p.}$ *has infinite multiplicity.*

Proof. Straight-forward.

* * * *

As one final topic related to spectral representations, let us define, Q(A), *the domain of a self-adjoint operator as a quadratic form:*

DEFINITION A.25. Let A be a self-adjoint operator. Then, in a spectral representation $Q(A) = \{\psi(x,n) \mid \int |x|\ |\psi(x,n)|^2\, d\mu < \infty$.

Remarks

1. Q(A) is the natural set for which $\langle\psi,A\psi\rangle$ makes sense.

2. For a fuller description of quadratic forms, even for non self-adjoint operators, see Kato [T17], Chap. VI.

THEOREM A.20. *Let* A *be bounded below, i.e.*, $E_{(-\infty,c)}(A) = 0$ *for some* C. *Then the quadratic form* $\psi \to \langle\psi,A\psi\rangle$ *defined on* $Q(A)$ is closed, i.e., *if* $\psi_n \in Q(A)$ *and* $\psi_n \xrightarrow{\|\ \|} \psi$, $\langle\psi_n, A\psi_n\rangle \to c$, *then* $\psi \in D(A)$ *and* $\langle\psi, A\psi\rangle = c$. *Moreover,* D(A) *is a core for the form in the sense that* A *on* Q(A) *is the smallest closed form extending* A *on* D(A).

Proof. Straight-forward.

(c). *One-Parameter Unitary Groups*

We collect here three crucial theorems on one parameter groups in Hilbert space.

DEFINITION A.26.

A family of unitary operators, U(t), obeying:

(a) $U(t)\,U(s) = U(t+s)$

(b) $t \to U(t)$ is strongly continuous

is called a *one parameter unitary group.*

THEOREM A.21 (Stone's theorem). *Let* U(t) *be a one parameter group. Let* $\mathfrak{D}_U = \{\psi \mid t^{-1}(U(t)-1)\psi$ *has a limit as* $t \to 0\}$. *Let* H *be defined with domain* $\mathfrak{D}_U$ *by:*

$$H\psi = -i \lim_{t \to 0} t^{-1}(U(t)-1)\psi.$$

Then H *is self-adjoint and* $U(t) = e^{iHt}$. *Conversely, let* H *be self-adjoint. Then* $U(t) = e^{iHt}$ *is a one parameter group with* $\mathfrak{D}_U = D(H)$.

Proof. See e.g., Nelson [T23], Section I.6 or Riesz-Nagy [T25], pp. 380-388.

Remark

Thus, we can paraphrase this theorem as "There is a one-one correspondence between one parameter groups and self-adjoint operators."

The next two theorems carry over naturally to any operator semi-groups in Banach spaces, which is their natural setting.

THEOREM A.22 (The Trotter-Kato theorem). *Let* A_n *be a family of self-adjoint operators and let* A *be self-adjoint. Let*

$$(A_n - z)^{-1} \xrightarrow{S} (A - z)^{-1}$$

for every z *with* $\operatorname{Im} z \neq 0$. *Then:*

$$e^{iA_n t} \xrightarrow{S} e^{iAt}$$

Proof. See Kato [T17], pp. 511-512 or Yoshida [T29], pp. 269-272. For the original proof, see Trotter [111]. ∎

THEOREM A.23 (The Trotter Product Formula). *Let* A *and* B *be self-adjoint with* A + B [*defined on* D(A) ∩ D(B)] *essentially self-adjoint. Then:*

$$\underset{n \to \infty}{\text{s-lim}} \left(e^{iAt/n} e^{iBt/n}\right)^n .$$

Proof. See Trotter [112] or Chernoff [20]. For a simple proof in case A + B is actually self-adjoint, see Appendix B of Nelson [91]. ∎

(d). *Compact Operators*

One of the most useful classes of operators in a Hilbert space is the class characterized by the following theorem:

THEOREM A.24. *The following characterizations of a bounded operator* T *are equivalent:*

(a) T *is the norm limit of operators of finite rank* (*i.e., operators of the form* $\sum_{n=1}^{N} |\psi_n\rangle \langle\phi_n|$).

(b) *If* $\phi_n \xrightarrow{W} \phi$, *then* $T\phi_n \xrightarrow{\|\ \|} T\phi$.

(c) *The image of the unit ball* $\{T|\ \|\Phi\| \leq 1\}$ *is compact (in* $\|\ \|$*-topology).*

Proof. See Riesz-Nagy [T25], pp. 178, 204, 206-208.

DEFINITION A.27. Any operator obeying one (and hence all) of the conditions of Theorem A.24 is called *compact.*

Compact operators have very nice spectral properties (even when they aren't self-adjoint).

THEOREM A.25 (the determinant free Fredholm theorems). *Let* A *be a compact operator. Then, there exists a set of points* $\lambda_1,\ldots,\lambda_n,\ldots$ *(finite or infinite) which either have no limit points or have only* 0 *as a limit point, so that:*

(a) *If* $\lambda \neq \lambda_i$ *all* i *and* $\lambda \neq 0$, *then* $(A-\lambda)^{-1}$ *exists as a bounded operator.*

(b) *If* $\lambda = \lambda_i$, *then* $A\phi = \lambda\phi$ *has a solution* $\phi \neq 0$.

Proof. See Riesz-Nagy [T25], pp. 160-172; 203-205.

One can extend this theorem to "analytic" families of operators. A priori, there are several possible definitions for analytic operators, but:

THEOREM A.26. *The following are equivalent conditions on an operator valued function* A(z), *defined for* z *in a domain* D *of the complex plane:*

(a) *For all* $\phi,\psi \in \mathcal{H}$, $\langle\phi, A(z)\psi\rangle$ *is analytic.*

(b) *For each* ϕ, $A(z)\phi$ *has a complex derivative [in the norm (vector) topology].*

(c) A(z) *has a complex derivative in the norm (operator) topology.*

(d) *For each* $z_0 \in D$, *there is an expansion converging for* z *near* z_0, *of the form:*

$$A(z) = \sum_{n=0}^{\infty} B_n(z-z_0)^n$$

where B_n *are bounded operators.*

Proof. See Kato [T17], p. 137.

The extension of the Fredholm theorem to analytic functions is:

THEOREM A.27 (the analytic Fredholm theorem). *Let* A(z) *be a compact operator-valued analytic function in a domain* D *of the complex plane. (By domain, we mean a connected open set.) Then, either:*

(a) $(1 - A(z))^{-1}$ *does not exist for any* $z \in D$.

(b) *There is a discrete set,* S, *in* D *so that* $(1 - A(z))^{-1}$ *exists if* $z \notin S$ *and* $A(z)\,\phi = \phi$ *has a solution of* $z \in S$. *Moreover* $(1 - A(z))^{-1}$ *is analytic in* D/S *and has poles at the points of* S.

Proof. See, for example, Hunziker [54], appendix; Tiktopoulos [109], appendix or Dunford-Schwartz, [T7]. ∎

Up to this point in our exposition, we have not described any really convenient criteria for compactness. A particularly useful criterion is:

THEOREM A.28 *Let* $\mathcal{H} = L^2(X, d\mu)$. *Let* A *have a kernel* A(x,y), i.e.:

$$(Af)(x) = \int_X A(x,y) f(y)\, d\mu(y).$$

If the kernel obeys $\int_{X x X} |A(x,y)|^2\, d\mu(x)\, d\mu(y) < \infty$, *then* A *is compact.*

Proof. Let $U_n(y)$ be an orthonormal basis for $L^2(X, d\mu)$. Then $U_n(x)\, U_n(y)$ is an orthonormal basis for $L^2(X x X,\, d\mu \otimes d\mu)$, so that, taking $a_{nm} = \int \overline{U_n(x)}\, U_m(y)\, A(x,y)$, we have:

$$\sum_{n,m<N} a_{nm}\, U_n(x)\, U_m(y) \xrightarrow{L^2} A(x,y)\,.$$

The left hand side is the kernel of the finite rank operator

$$\sum_{n,m<N} a_{nm}\, |U_n\rangle \langle U_m|\,.$$

Since $\int d\mu(x) | \int d\mu(y)\, B(x,y) f(y)|^2 \leq \int d\mu(x)\, d\mu(y)\, |B(x,y)|^2 \times \int d\mu(y)\, |f(y)|^2$, the $L^2\,(X \times X)$ convergence implies $\|\ \|$ convergence. Thus A is the norm limit of finite rank operators. ∎

DEFINITION A.28. An operator A with kernel A(x,y) obeying $\int |A(x,y)|^2$ obeying $\int |A(x,y)|^2\, d\mu(x)\, d\mu(y) < \infty$ is called *Hilbert-Schmidt* (H.-S.).

There is a close relation between H.-S. operators and the trace.

DEFINITION A.29. We say $A \in \mathcal{J}_1$ *(trace class)* if and only if, for every orthonormal basis, $\{\phi_n\}$:

$$\sum_n |\langle \phi_n, A\phi_n \rangle| < \infty.$$

We say $A \in \mathcal{J}_2$ (abstract H.-S.) if and only if for every orthonormal basis, $\{\phi_n\}$:

$$\sum_n \|A\phi_n\|^2 < \infty.$$

The properties of these two classes are summarized in:

THEOREM A.29.

(a) $\mathcal{J}_1$ *and* $\mathcal{J}_2$ *are* * *ideals, i.e., they are closed under sums, adjoints, and multiplication by bounded operators.*

(b) $A \in \mathcal{J}_1$ *if and only if* $A = BC$ *with* $B, C \in \mathcal{J}_2$ [i.e., $[\mathcal{J}_2]^2 = \mathcal{J}_1$].

(c) $\mathcal{J}_1 \subset \mathcal{J}_2 \subset$ *Compact Operators.*

(d) *For any* A, $A \in \mathcal{J}_2$ *if* $\sum_n \|A\phi_n\|^2 < \infty$ *for a single basis. For* $A \geq 0$, $A \in \mathcal{J}_1$ *if* $\sum \langle \phi_n, A\phi_n \rangle < \infty$ *for a single basis.*

(e) *If* $A \in \mathcal{J}_1$, $\mathrm{Tr}(A) = \sum \langle \phi_n, A\phi_n \rangle$ *is independent of the basis used.*

(f) Tr *is linear and obeys* $\mathrm{Tr}(A^*) = \overline{\mathrm{Tr}(A)}$; $\mathrm{Tr}(AB) = \mathrm{Tr}(BA)$ [*if* $A \in \mathcal{J}_1$, B *bounded or if* $A, B \in \mathcal{J}_2$].

(g) $\mathcal{J}_2$ *endowed with the inner product* $\langle A,B \rangle_2 = \mathrm{Tr}(A^*B)$ *is a Hilbert space. If* $\|A\|_2 = [\mathrm{Tr}(A^*A)]^{1/2}$, *then* $\|A\|_2 \geq \|A\|$ *and* $\mathcal{J}_2$ *is the* $\| \ \|_2$*-closure of the finite rank operators.*

(h) $\mathcal{J}_1$ *endowed with the norm* $\|A\|_1 = \mathrm{Tr}(\sqrt{A^*A})$ *is a Banach space.* $\|A\|_1 \geq \|A\|_2 \geq \|A\|$ *and* $\mathcal{J}_1$ *is the* $\| \ \|_1$*-norm closure of the finite rank operators. The dual space of* $\mathcal{J}_1$ *is* $\mathcal{B}(\mathcal{H})$, *the family of bounded operators with the operator norm and with the duality* $\langle B,A \rangle = \mathrm{Tr}(BA)$.

(i) *If* A, B $\in \mathcal{I}_2$, *then* $\|AB\|_1 \leq \|A\|_2 \|B\|_2$. *If* A $\in \mathcal{I}_2$, B $\in \mathcal{B}(\mathcal{H})$, *then* $\|AB\|_2 \leq \|A\|_2 \|B\|$. *If* A $\in \mathcal{I}_1$, B $\in \mathcal{B}(\mathcal{H})$, *then* $\|AB\|_1 \leq \|A\|_1 \|B\|$.

Proof. See Kato [T17], pp. 262-264, 519-521. ∎

Finally, let us look at the connection between abstract Hilbert-Schmidt operators and Hilbert-Schmidt operators as defined by kernel conditions. As the names imply:

THEOREM A.30. *Let* $\mathcal{H} = L^2(X, d\mu)$. *Then any operator* A, *with Hilbert-Schmidt kernel has* A $\in \mathcal{I}_2$. *Conversely, if* A $\in \mathcal{I}_2$, *then* A *has a Hilbert-Schmidt kernel* A(x,y). *Moreover* $\int |A(x,y)|^2\, dxdy = \|A\|_2^2$ $(\equiv \mathrm{Tr}(A^*A))$.

Proof. Let A have H.S. kernel. Let $U_n(x)$ be an orthonormal basis for $L^2(X,d\mu)$. Then:

$$\Sigma \|AU_n\|^2 = \sum_{n,m} |\langle U_n, AU_m\rangle|^2$$

$$= \sum_{n,m} \left| \int \overline{U_n(x)} A(x,y)\, U_m(y)\, d\mu(x)\, d\mu(y)\right|^2$$

$$= \sum_{n,m} |a_{nm}|^2 = \int |A(x,y)|^2\, d\mu(x)\, d\mu(y)$$

since $\overline{U_n(x)}\, U_m(y)$ is an O.N. basis. Conversely, let A $\in \mathcal{I}_2$ and let $a_{nm} = \langle U_n, AU_m\rangle$. Then $\|A\|_2^2 = \Sigma\, |a_{nm}|^2 < \infty$ so

$$\sum_{n<N\ m<N} a_{nm}\, U_n(x)\, \overline{U_m(y)}$$

converges in $L^2(X \times X, d\mu \quad d\mu)$ norm to an H.-S. kernel A(x,y) which yields an operator $\tilde{A}$. Since $\langle U_n, \tilde{A}U_m\rangle = a_{nm} = \langle U_n, AU_m\rangle$ we have $A = \tilde{A}$. ∎

(e) *The Regularity Theorem for* Δ

THEOREM A.31. *Let* V *be a* C^∞ *function in a region,* D, *of* $\mathbb{R}^3$. *Let* f *be a weak* L^2 *solution of* $-\Delta f + Vf = 0$ *in* D, *i.e.,* f *is locally* L^2 *and*

$\langle f,(-\Delta + V)g\rangle = 0$ *for all functions* g *which are* C^∞ *and have compact support in* D. *Then* f *is* C^∞ *in* D.

Proof. See, e.g., Hörmander's *Linear Partial Differential Operators.* ∎

Remarks

This is a special case of a very general result about elliptic partial differential equations. Hörmander treats the general case. Using the fact the $-\Delta$ has constant (i.e., non x-dependent) coefficients, one can construct a much simpler proof.

REFERENCES

A. *Articles, Theses, Summer School Notes*

[1] Aaron R. and Klein, A.: "Convergence of the Born Expansion." J. Math. Phys. 1 (1960), 131-138.

[2] Aronszajn, N.: "On a problem of Weyl in the theory of singular Sturm Liouville equations." Am. J. Math. 79 (1957), 597-610.

[3] Babbitt, D.: "The Weiner integral and perturbation theory of the Schrödinger operator." Bull. Am. Math. Soc. 70 (1964), 254-259.

[4] Bargmann, V.: "On the number of bound states in a central field of force." Proc. Nat. Acad. Sci. (U.S.A.) 38 (1952), 961-966.

[5] ______________: "On unitary ray representations of continuous groups." Ann. Math. 59 (1954), 1-62.

[6] ______________: "A note on Wigner's theorem on symmetry operations." J. Math. Phys. 5 (1964), 862-868.

Bertero, M: see [134].

[7] Birman, M: "Conditions for the existence of wave operators." (Russian). Dokl. Akad. Nauk. SSSR 147 (1962), 1008-1009. [English trans. = Soviet Math. Dokl. 3 (1962), 408-411].

Brander, O.: see [136].

[8] Brownell, F. H.: "A note on Cook's wave matrix theorem." Pac. J. Math. 12 (1962), 47-52.

[9] Brenig, W. and Haag, R.: "General quantum theory of collision processes." [Eng. trans = Reprinted in *Quantum Scattering Theory.* ed. M. Ross; Indiana Univ. Press, (1963, 13-108].

[10] Calderón, A. P.: "Intermediate spaces and interpolation." Studia Math. *Seria Spec. 1,* (1960), 31-34.

[11] Calkin, J.: "Symmetric transformations in Hilbert space." Duke J. Math. 7 (1940), 504-508.

[12] Calogero, F.: "Sufficient conditions for an attractive potential to possess bound states I, II." J. Math. Phys. 6 (1965), 161-164, 1105-1107.

[13] ____________: "Necessary condition for the existence of bound states." Il Nuovo Cimento 36 (1965), 199-201.

[14] ____________ : "Upper and lower limits for the number of bound states in a given central potential." Commun. Math. Phys. 1 (1965), 80-88.

[15] Cameron, R.: "The Ilstow and Feynman integrals." J. Anal Math. 10 (1962/63), 287-361.

[16] Cannon, J. T.: "Field Theoretic Properties of Nelson's Model." Princeton Thesis (1968), unpubl.

[17] Carleman, T.: "Sur la théorie mathematique de l'equation de Schrödinger." Arkiv för Mat., Ast. & Fys. 24B No. 11 (1934).

[18] Case, K. M.: "Singular potentials." Phys. Rev. 80, (1950), 797-806.

[19] Chadan, K.: "The asymptotic behavior of the number of bound states of a given potential in the limit of large coupling." Il Nuovo Cimento 58 (1968), 191-204.

[20] Chernoff, P.: "Note on product formulas for operator semigroups." J. Func. Anal. 2 (1968), 238-242.

[21] Coester, F.: "Single channel scattering amplitude." Phys. Rev. 133 (1964), B1516-B1519.

[22] Cohn, J. H. E.: "On the number of negative eigenvalues of a singular boundary value problem." J. London Math. Soc. 40 (1965), 523-525.

[23] __________: "On the number of bound states in a central field of force." J. London Math. Soc. 41 (1966), 474-478.

Combes, J. M.: see [145].

[24] Cook, J. M.: "Convergence of the Møller wave-matrix." J. Math. and Phys. 36 (1957), 82-87.

[25] Daletzski, Yu.: "Continuous integrals associated with operator evolution equations." (Russian), Usp. Mat. Nauk. 17 (1962), 3-115.

[26] Dollard, J.: "Non-relativistic time-dependent scattering theory and the Coulomb interaction." Princeton thesis (1963), unpubl.

[27] __________: "Asymptotic convergence and the Coulomb interaction." J. Math. Phys. 5 (1964), 729-738.

[28] Eberlein, W.: "Closure, convexity and linearity in Banach spaces." Ann. Math. 47 (1946), 688-703; esp. p. 699.

Eckart, C.: see [138].

[29] Faris, W.: "The product formula for semigroups defined by Freidrichs' extensions." Pac. J. Math. 22 (1967), 47-70.

[30] __________: "Product formulas for perturbations of linear propagators." J. Func. Anal. 1 (1967), 93-108.

[31] Fatou, P.: "Séries trigonométriques et séries de Taylor." Acta. Math. 30 (1906), 335-400.

[32] Feldman, J.: "On the Schrödinger and heat equations for non-negative potentials." Trans. Am. Math. Soc. 108 (1963), 251-264.

[33] Feynman, R. P.: "Space-time approach to non-relatavistic quantum mechanics." Rev. Mod. Phys. **20** (1948), 367-387.

[34] Fonda, L. and Ghirardi, G. C.: "Approximate determination of the number and energies of bound states of a physical system." Il Nuovo Cimento **46** (1966), 47-58.

[35] Frank, W.: "Strong Coupling Limit in Potential Theory, I." J. Math. **8** (1967), 466-472.

[36] Freudenthal, H.: "Über die Friedrichssche Fortsetzung halbbeschränkter Hermitescher Operatoren." Proc. Acad. Amsterdam **39** (1936), 832-833.

[37] Friedrichs, K.: "Spectraltheorie halbbeschränkter Operatoren I-III." Math. Ann. **109** (1934), 465-487, 685-713; **110** (1935), 777-779.

[38] ____________ : "Über die Spektralzerlegung eines Integraloperators." Math. Ann. **115** (1938), 249-272.

[39] ____________: "On the perturbation of continuous spectra." Comm. Pure Appl. Math. **1** (1948), 361-406.

[40] Gel'fand, I. M. and Yaglom, A. M.: "Integration in function spaces and its applications to quantum physics." J. Math. Phys. **1** (1960), 48-69.

[41] Gell-Mann, M. and Goldberger, M. L.: "The formal theory of scattering." Phys. Rev. **91** (1953), 398-408.

[42] Ghirardi, G. C. and Rimini, A.: "On the number of bound states of a given interaction." J. Math. Phys. **6** (1965), 40-44.

____________: see also, Fonda, L. and ____________.

Goldberger, M. L.: see Gell-Mann, M. and ____________.

[43] Green, T. A. and Lanford, O. E. III: "Rigorous derivation of the phase shift formula for the Hilbert space scattering operator of a single particle." J. Math. Phys. 1 (1960), 131-140.

[44] Grossman, A. and Wu, T.T.: "Schrödinger Scattering Amplitude, I." J. Math. Phys. 2 (1961), 710-713.

[45] ____________ and ____________: "Schrödinger Scattering Amplitude, III." J. Math. Phys. 3 (1962), 684-689.

Haag, R.: see Brenig, W. and ____________.

[46] Hack, M. N.: "Bound states and the formal theory of scattering." Phys. Rev. 96 (1954), 196-198.

[47] ____________: "On convergence to the Møller wave operators." Il Nuovo Cimento 9 (1958), 731-733.

[48] Hardy, G. H. and Littlewood, J. E.: "Some properties of fractional integrals (1)." Math. Zeit. 27 (1928), 565-608.

[49] Hepp, K.: "On the quantum mechanical N-body problem." Helv. Phys. Acta 42 (1969), 425-459.

Hughes, D. S.: see [138].

[50] Hunt, R.: "An extension of the Macinkiewicz interpolation theorem to Lorentz spaces." Bull. Am. Math. Soc. 70 (1964), 803-807.

[51] ____________: "On L(p,q) spaces." L'Enseignment Math. 12 (1966), 250-276.

[52] Hunziker, W.: "Regularitätseigenschaften der Streuamplitude im Fall der Potentialstreuung." Helv. Phys. Acta 34 (1961), 593-620.

[53] ____________: "Proof of a conjecture of S. Weinberg." Phys. Rev. 135 (1964), B800-B803.

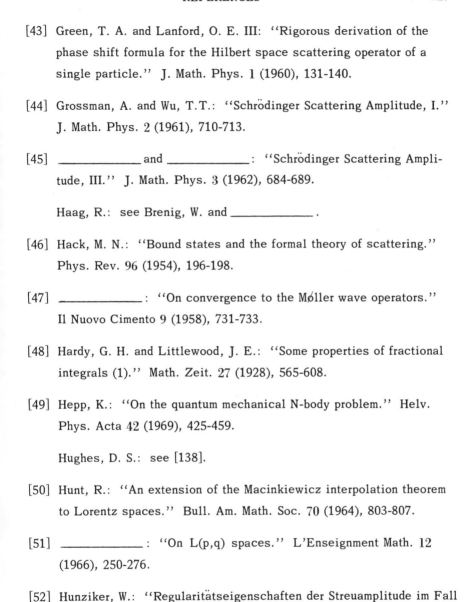

[54] ____________: "On the spectra of Schrödinger multiparticle Hamiltonians." Helv. Phys. Acta 39 (1966), 451-462.

[55] ____________: "Mathematical theory of multiparticle quantum systems." In *Lectures in Theoretical Physics*, Vol. X-A. Ed. A. O. Barut and W. Britten; Gordon and Breach, New York, 1968 (1967 Colorado summer school lectures).

[56] Ikebe, T.: "Eigenfunction expansions associated with the Schrödinger operators and their applications to scattering theory." Arch. Rat. Mech. Anal. 5 (1960, 1-34.

See also [142], [144].

[57] ____________ and Kato, T.: "Uniqueness of self-adjoint extensions of singular elliptic differential operators." Arch. Rat. Mech. Anal. 9 (1962), 77-92.

[58] Ito, K.: "Weiner integrals and Feynman integrals." in *Proc. 4th Berkeley Symposium on Mathematics, Statistics and Probability, Vol. II.*, pp. 228-238. Univ. of Calif. Press, Berkeley, 1961.

[59] Jaffe, A.: "Dynamics of a Cut-off $\lambda\phi^4$ Field Theory." Princeton Thesis (1965), unpubl.

[60] ____________: A.M.S. Memoir, to appear.

[61] Jauch, J. M.: "Theory of the scattering operator, I, II." Helv. Phys. Acta. 31 (1958), 127-158, 661-684.

[62] ____________ and Zinnes, I. I.: "The asymptotic condition for simple scattering systems." Il Nuovo Cimento 11 (1959), 553-567.

[63] Jost, R. and Pais, A.: "On the scattering of a particle by a static potential." Phys. Rev. 82 (1951), 840-851.

[64] Kac, M.: "On some connections between probability theory and differential and integral equations." *Proc. 2nd Berkeley Symposium on Mathematics and Probability,* pp. 189-215. Univ. of Calif. Press, Berkeley, 1950.

[65] Kato, T.: "Fundamental properties of Hamiltonian operators of Schrödinger type." Trans. Am. Math. Soc. 70 (1951), 195-211.

[66] ____________: "Integration of the equation of evolution in a Banach space." J. Math. Soc. Japan 5 (1953), 208-234.

[67] ____________: "On finite dimensional perturbations of self-adjoint operators." J. Math. Soc. Japan 9 (1957), 239-249.

[68] ____________: "Perturbation of continuous spectra by trace class operators." Proc. Japan Acad. 33 (1957), 260-264.

[69] ____________: "Growth properties of solutions of the reduced wave equation with variable coefficient." Comm. Pure Appl. Math. 12 (1959), 403-425.

[70] ____________: "Wave operators and similarity for non self-adjoint operators." Math. Ann. 162 (1966), 258-279.

[71] ____________: "Some mathematical problems in quantum mechanics." Supp. Prog. Theor. Phys. 40 (1967), 3-19.

[72] ____________: "Some results in potential scattering." In *Proceedings of International Conference on Functional Analysis and Related Topics.* 1969

[73] ____________ and Kuroda, S. T.: "A remark on the unitarity property of the scattering operator." Il Nuovo Cimento 14 (1959), 1102-1107.

[74] ____________ and ____________: "Theory of simple scattering and eigenfunction expansions." In *Functional Analysis and Related Fields.* Springer, 1969

____________: see also Ikebe, T., and ____________, and [139].

[75] Khuri, N.: "Analyticity of the Schrödinger scattering amplitude and non-relativistic dispersion relations." Phys. Rev. 107 (1957), 1148-1156.

Kinoshita, T.: see [137].

[76] Kisyński, J.: "Sur les opérateurs de Green des problèmes de Cauchy abstraits." Studia Mathematica 23 (1964), 285-328.

Klein, A.: see Aaron, R. and ____________; and Zemach, C. and ____________.

[77] Kunze, R. A. and Stein, E. M.: "Uniformly bounded representations and harmonic analysis of the 2 x 2 real unimodular group." Am. J. Math. 82 (1960), 1-62.

[78] Kupsch, J. and Sandhas, W.: "Møller operators for scattering on singular potentials." Commun. Math. Phys. 2 (1966), 147-154.

[79] Kuroda, S. T.: "On the existence and the unitarity property of the scattering operator." Il Nuovo Cimento 12 (1959), 431-454.

[80] ____________: "Stationary methods in the theory of Scattering." In *Perturbation Theory and its Applications in Quantum Mechanics*. Ed. C. Wilcox, pp. 185-214. Wiley, New York, 1966.

[81] ____________: "Stationary theory of scattering and eigenfunction expansions I, II." (Japanese), Sûgaku 18 (1966), 74-85, 137-144.

[82] ____________: "Perturbation of eigenfunction expansions." Proc. Nat. Acad. Sci. (U.S.A.), 57 (1967), 1213-1217.

[83] ____________: "An abstract stationary approach to perturbation of continuous spectra and scattering theory." J. Anal. Math. 20 (1967), 57-117.

[84] Ladyzhenskaya, O. A., and Visik, I. M.: "Problèmes aux limites pour les equations aux dérivées partielles et certaines classes d'équations operationelles." (Russian), Usp. Mat. Nauk **11** (1956), 41-97.

Lanford, O. E. III: see Green T. A., and ___________.

Limić, N.: see [131].

[85] Lions, J. L., and Peetre, J.: "Sur une classe d'espaces d'interpolation." I.H.E.S. Publ. Math. No. **19** (1964), 5-68.

[86] Lippmann, B. A., and Schwinger, J.: "Variational principles for scattering processes, I." Phys. Rev. **79** (1950), 469-480.

Littlewood, J. E.: see Hardy, G. H., and ___________.

Loeffel, J: see [137].

[87] McShane, E.: "Integrals devised for special purposes." Bull. A.M.S. **69** (1963), 597-627.

[88] Marcinkiewicz, J.: "Sur l'interpolation d'operations." C. R. Acad. Sci. Paris **208** (1939), 1272-1273.

Martin, A.: see [135, 137].

[89] Møller, C.: "General properties of the characteristic matrix in the theory of elementary particles." Det. Kgl. Dansk. Vid. Selsk. Mat.-Fys. Medd. **23** No. 1 (1945).

[90] Moses, H. E.: "The scattering operator and the adiabatic theorem." Il Nuovo Cimento **1** (1955), 104-131.

[91] Nelson, E.: "Feynman integrals and the Schrödinger equation." J. Math. Phys. **5** (1964), 332-343.

___________: see also [132].

[92] von Neumann, J.: "Allgemeine Eigenwerttheorie Hermitescher Functional-operatoren." Math. Ann. **102** (1929), 49-131; esp. 103.

[93] ____________ and Wigner, E. P.: "Uber merkwurdige diskrete Eigenwerte." Z. Physik 30 (1929), 465-467.

[94] Odeh, F.: "Note on differential operators with purely continuous spectrum." Proc. Am. Math. Soc. 16 (1965), 363-366.

Pais, A.: see Jost, R., and ____________.

Peetre, J.: see Lions, J. L., and ____________.

[95] du Plessis, N.: "Some theorems about the Reisz fractional integral." Trans. Am. Math. Soc. 80 (1955), 124-134.

[96] Putnam, C. R.: "Continuous spectra and unitary equivalence." Pac. J. Math. 7 (1957), 993-995.

[97] Rellich, F.: "Störungstheorie der Spektralzerlegung I-V." Math. Ann. 113 (1937), 600-619, 677-685; 116 (1939), 555-570; 117 (1940), 356-382; 118 (1942), 462-484.

[98] Rollnik, H.: "Streumaxima und gebundene Zustände." Z. Phys. 145 (1956), 639-653.

[99] Rosenblum, M.: "Perturbation of the continuous spectrum and unitary equivalence." Pac. J. Math. 7 (1957), 997-1010.

Sandhas, W.: see Kupsch, J. and ____________.

[100] Scadron, M., Weinberg, S., and Wright, J.: "Functional analysis and scattering theory." Phys. Rev. 135 (1964), B202-B207.

[101] Schwinger, J.: "On the bound states of a given potential." Proc. Nat. Acad. Sci. (U.S.A.), 47 (1961), 122-129.

____________: see also Lippmann, B. A., and ____________.

Sigalov, A. G.: see [141].

[102] Simon, B.: "On the growth of the number of bound states with increase in potential strength." J. Math. Phys. 10 (1969), 1123-1126.

[103] ______________ : "On positive eigenvalues of one-body Schrödinger operators." Comm. Pure Appl. Math 22 (1969), 531-538.

______________: see also [143].

[104] Sobolev, S. L.: "Sur un théorème d'analyse fonctionnelle." (Russian), Math. Sbornik. (Recueil Math.) 46 (N.S.4) (1938), 471-496.

[105] Stein, E. M., and Weiss, G.: "An extension of a theorem of Marcinkiewicz and some of its applications." J. Math. Mech. 8 (1959), 263-284.

______________: see also Kunze, R. A., and ______________.

[106] Strichartz, R.: Multipliers on Generalized Sobolev Spaces. Princeton Thesis (1966), unpubl.

[107] ______________: "Multipliers on fractional Sobolev spaces. J. Math. Mech. 16 (1967), 1031-1060.

Talenti, G.: see [134].

[108] Thoe, D. W.: "Eigenfunction expansions associated with Schrödinger operators in R_n, n > 4." Arch. Rat. Mech. and Anal. 26 (1967), 335-356.

[109] Tiktopoulos, G.: "Analytic continuation in complex angular momentum and integral equations." Phys. Rev. 133 (1964), B1231-1238.

[110] ______________ : Lectures for Physics 575, (Fall 1968), unpubl.

[111] Trotter, H. F.: "Approximation of semi-groups of operators." Pac. J. Math. 8 (1958), 887-919.

[112] ______________: "On the product of semi-groups of operators." Proc. Am. Math. Soc. 10 (1959), 545-551.

Visik, I. M.: see Ladyzhenskaya, O. A., and ______________.

Viano, G. A.: see [134].

[113] Weidmann, J.: "On the continuous spectrum of Schrödinger operators." Comm. Pure Appl. Math. 19 (1966), 107-110.

[114] ____________: "The Virial theorem and its applications to the spectral theory of Schrödinger operators." Bull. Amm. Math. Soc. 73 (1967), 452-456.

[115] ____________: "Zur Spektral theorie von Sturm-Liouville-Operatoren." Math. Zeit. 98 (1967), 268-302.

[116] Weinberg, S.: "Quasiparticles and the Born series." Phys. Rev. 131 (1963), 440-460.

[117] ____________: "Systematic solution of multiparticle scattering problems." Phys. Rev. 133 (1964), 232-256.

[118] ____________: "Quasiparticles and perturbation theory." In *Lectures on Particles and Field Theory*. Ed. S. Deser and K. Ford, pp. 289-403; Prentice-Hall, Englewood Cliffs, N. J., 1965. (1964 Brandeis Lectures, Vol. II).

____________: see also Scandron, M., ____________ and Wright, J.

Weiss, G.: see Stein, E. M., and ____________.

[119] Weyl, H.: "Über gewöhnliche (lineare) Differentialgleichungen mit singulären Stellen and ihre Eigenfunctionen." Nachr. Akad. Wiss. Gott. (1909), 37-64; (1910), 442-467.

[120] ____________: "Über gewöhnliche Differentialgleichungen mit Singularitäten und die zugehorigen Entwicklungen willkürlicher Funktionen." Math. Ann. 68 (1910), 220-269.

[121] Wightman, A. S.: "Relatavistic invariance and quantum mechanics." Supp. Nuovo Cimento 14 (1959), 81-94.

[122] ____________: "Introduction to some aspects of the relatavistic dynamics of quantized fields." In *1964 Cargèse Lectures*. Ed. Levy, pp. 171-291. Gordon and Breach, New York, 1967.

[123] Wigner, E. P.: "On unitary representations of the inhomogeneous Lorentz group." Ann. Math. 40 (1939), 149-204, esp. pp. 169-170.

____________: see also von Neumann, J., and ____________.

[124] van Winter, C.: "Theory of finite systems of particles, I, II." Mat. Fys. Skr. Dan. Vid Selsk 1 Nos. 8, 10 (1964/65).

[125] Wintner, A.: "On the normalization of characteristic differentials in continuous spectra." Phys. Rev. 72 (1947), 516-517.

Wright, J.: see Scandron, M., Weinberg, S., and ____________.

Wu, T. T.: see Grossman, A., and ____________.

Yaglom, A.: see Gel'fand, I. M., and ____________.

[126] Yoshida, K.: "Time dependent evolution equations in a locally compact space." Math. Ann. 162 (1966), 83-86.

[127] ____________: "On holomorphic Markov processes." Proc. Japan Acad. 42 (1966), 313-317.

[128] ____________: "On the integration of the equation of evolution." J. Fac. Sci. Univ. Tokyo, Sect. 1, 9, Part 5, (1963), 397-402.

[129] Zemach, C., and Klein, A.: "The Born expansion in non-relatavistic quantum theory." Il Nuovo Cimento 10 (1958), 1078-1087.

Zhislin, G. M.: see [140, 141].

Zinnes, I. I.: see Jauch, J. M., and ____________.

[130] Zygmund, A.: "On a theorem of Marcinkiewicz concerning interpolation of operators." J. Math. Pure Appl. 35 (1956), 223-248.

[131] Limić, N.: "On the Self-Adjointness of the Operator $-\Delta + V$." Commun Math. Phys. 1 (1966), 321-327.

[132] Nelson, E.: "Interaction of nonrelatavistic particles with a quantized scalar field." J. Math Phys. 5 (1964), 1190-1197.

[133] Jörgens, K.: "Wesentliche Selbstadjungiertheit Singulärer Elliptischer Differentialoperatoren Zweiter Ordnung in $C_0^\infty(G)$." Math. Scand. 15 (1964), 5-17.

[134] Bertero, M., Talenti, G., and Viano, G.A.: "Eigenfunction expansions associated with Schrödinger two-particle operators." Nuovo Cimento 62 (1969), 27-87.

[135] Martin, A.: "Born approximation and dispersion relations for singular potentials." In *Preludes in Theoretical Physics in Honor of V. Weisskopf.* Ed. De-Shalit, Feshbach, Van Hove. North Holland, Amsterdam, 1966.

[136] Brander, O.: "Proof of the Mandelstam representation for logarithmically singular potentials." Nuoco Cimento 61 (1969), 605-636.

[137] Kinoshita, T.; Loeffel, J. J., and Martin, A.: "Upper Bounds for the Scattering Amplitude at High Energy." Phys. Rev. 135 (1964), B1464-1482.

[138] D. S. Hughes and C. Eckart, "The Effect of the Motion of the Nucleus on the Spectra of Li I and Li II." Phys. Rev. 36 (1930), 694-698.

[139] Kato, T.: "On the Existence of Solutions of the Helium Wave Equation." Trans. A.M.S. 70 (1951), 212-218.

[140] Zhislin, G. M., "Discussion of the Spectrum of the Schrödinger Operator for Systems of Many Particles" (in Russian), Tr. Moskovsk. matem. obsshch. 9 (1960), 81-128.

[141] ______________ and Sigalov, A. G.: "Mixed Spectrum of certain Multidimensional Quantum Mechanical Differential Operators." Soviet Physics "Doklady", 9 (1965, 648-649. [Russian Original: Dok. Akad. Nauk SSSR 157 (1964), 1329-1331.]

[142] Ikebe, T.: "On the Phase-Shift formula for the Scattering Operator." Pac. J. Math. 15 (1965), 511-523.

[143] Simon, B.: "On the Infinitude or Finiteness of the Number of Bound States for an N-body Quantum System." Helv. Phys. Acta. 43 (1970), 607-630.

[144] Ikebe, T.: "Orthogonality of the Eigenfunctions for the Exterior Problem connected with $-\Delta$." Arch. Rat. Mech. Anal. 19 (1965), 71-73.

[145] Combes, J. M.: "Relatively Compact Interactions in Many Particle Systems." Commun. Math. Phys. 12 (1969), 283-295.

[146] Ciafaloni, M., and Menotti, P.: "Analysis of S-Matrix Singularities by Means of Operator Techniques." Nuovo Cimento 35 (1965), 160-192.

B. *Texts and Monographs*

[T1] de Alfaro, V., and Regge, T.: *Potential Scattering.* North Holland Co., Amsterdam, 1965.

[T2] Berberian, S.: *Measure and Integration.* MacMillan, New York, 1965.

[T3] Bers, L., John, F., and Schecter, M.: *Partial Differential Equations.* Interscience, New York, 1964.

[T4] Calogero, F.: *Variable Phase Approach to Potential Scattering.* Academic Press, New York, 1967.

[T5] Coddington, E., and Levinson, N.: *Theory of Ordinary Differential Equations.* McGraw-Hill, New York, 1955.

[T6] Courant, R., and Hilbert, D.: *Methods of Mathematical Physics,* Vol. I. Interscience, New York, 1953.

[T7] Dunford, N., and Schwartz, J.: *Linear Operators*, Part II. Interscience, New York, 1963.

[T8] Feynman, R. P., and Hibbs, A. R.: *Quantum Mechanics and Path Integrals*. McGraw Hill, New York, 1965.

[T9] Goldberger, M. L., and Watson, K.: *Collision Theory*. Wiley, New York, 1964.

[T10] Gradshteyn, I. S., and Ryzhik, I. W.: *Table of Integrals, Series and Products*. Academic Press, New York, 1965.

[T11] Halmos, P. R.: *Measure Theory*. D. Van Nostrand Co., Princeton, N. J., 1950.

[T12] ____________: *Introduction to Hilbert Space and the Theory of Spectral Multiplicity*. Chelsea, New York, 1951.

[T13] Hardy, G. H.: *Divirgent Series*. Oxford Univ. Press, London, 1949.

[T14] ____________, Littlewood, J. E., and Pólya, G.: *Inequalities*, 2nd ed. Cambridge

Hibbs, A. R.: see Feynman, R. P., and ____________.

Hilbert, D.: see Courant, R., and ____________.

[T15] Hoffman, K.: *Banach Spaces of Analytic Functions*. Prentice Hall, Englewood Cliffs, N. J., 1962.

John, F.: see Bers, L., ____________, and Schecter, M.

[T16] Kac, M.: *Probability and Related Topics in the Physical Sciences*. Interscience, London, 1959.

[T17] Kato, T.: *Perturbation Theory for Linear Operators*. Springer-Verlag, New York, 1966.

[T18] Landau, L., and Lifschitz, E.: *Quantum Mechanics, Non-Relatavistic Theory*. Pergamon, London, 1958.

Levinson, N.: see Coddington, E. A., and ______________.

Lifschitz, E.: see Landau, L., and ______________.

[T19] Lions, J. L.: *Equations differentielles operationelles.* Springer-Verlag, Berlin, 1961.

Littlewood, J. E.: see Hardy, G. H., ______________ and Pólya, G.

Sz. Nagy: see Reisz, F., and ______________.

[T20] Lorch, E.: *Spectral Theory.* Oxford Univ. Press, New York, 1962.

[T21] Mackey, G. W.: *Mathematical Foundations of Quantum Mechanics.* Benjamin, New York, 1963.

[T22] Nelson, E.: "Operator Differential Equations." Notes for Math 520 (Princeton Univ. 1964/65; notes by J. Cannon) unpubl.

[T23] ______________: *Topics in Dynamics I: Flows.* Princeton Univ. Press, 1970.

[T24] Newton, R.: *Scattering Theory of Waves and Particles.* McGraw-Hill; New York, 1966.

Polya, G.: see Hardy, G. H., Littlewood, J. E., and ______________.

Regge, T.: see de Alfaro, V., and ______________.

[T25] Riesz, F., and Sz-Nagy, B.: *Functional Analysis.* Ungar, New York 1955.

Ryzhik, I. W.: see Gradshteyn, I. S., and ______________.

Schector, M.: see Bers, L., John, F., and ______________.

Schwartz, J.: see Dunford, N., and ______________.

[T26] Stone, M. H.: *Linear Transformations in Hilbert Space and Their Applications to Analysis.* Colloq. Publ. A.M.S., New York, 1932.

Watson, G. N.: see Whittaker, E. T., and ____________.

Watson, K.: see Goldberger, M. L., and ____________.

[T27] Whittaker, E. T., and Watson, G. N.: *A Course in Modern Analysis*. Cambridge Univ. Press, Cambridge, 1965 (4th edition).

[T28] Wigner, E. P.: *Group Theory*. Academic Press, New York, 1959.

[T29] Yoshida, K.: *Functional Analysis*. Academic Press, New York, 1965.

[T30] Zygmund, A.: *Trigonometric Series, II*. Cambridge Univ. Press, London, 1959.

LIST OF SYMBOLS

INDEX